Nanotechnology in Industrial Wastewater Treatment

Nanotechnology in Industrial Wastewater Treatment

Dr. Arup Roy and Prof. Jayanta Bhattacharya

IWA PUBLISHING

Published by

IWA Publishing
Alliance House
12 Caxton Street
London SW1H 0QS, UK
Telephone: +44 (0)20 7654 5500
Fax: +44 (0)20 7654 5555
Email: publications@iwap.co.uk
Web: www.iwapublishing.com

First published 2015

British Library Cataloguing in Publication Data
A CIP catalogue record for this book is available from the British Library

ISBN 9781780406879 (Paperback)
ISBN 9781780406886 (eBook)

Contents

List of Figures

List of Tables

About the Authors

Arup Roy

Arup Roy has Bachelor's and Master's degrees in Microbiology and Ph.D. in Engineering with specialisation in waste water treatment. His research deals with synthesis and characterization of nanomaterials and their application in waste water treatment as well as in other extramural areas. He is also trained in EIA and EMP preparation. He is presently an Assistant Leader – EIA at Hubert Enviro Care Systems (P) Ltd, India.

Jayanta Bhattacharya

Jayanta Bhattacharya, PhD is Professor and Head, School of Environmental Science and Engineering and Professor in the Department of Mining Engineering, Indian Institute of Technology, Kharagpur, India. He is also a Visiting Professor at Western Australia School of Mines, Kalgoorlie, under Curtin University. He has made significant contributions to the mining academics by authoring books on mining that are taught all over the world.

Prof. Bhattacharya is a famous scientist in the area of environmental technology and management. He and his team developed a chemo-bioreactor for wastewater bioremediation, a community Gravity Filter – among few others. Recently he and his students made pioneering contribution in the development of CaO, $Ca(OH)_2$ and CaS nanoparticles for rapid precipitation of in-soluble compounds from wastewater streams. These particles would find applications as catalyst, toxic waste remediation agent and absorbent – to name a few. He and his team developed and established a novel spectro-photometric method to measure Sulfate concentrations in the acidic/mining wastewater without Interference of mainly, arsenate and phosphate ions. A continuous theme of research by his team is to develop new technologies for natural Carbon-Hydrogen-Nitrogen dosage supplements and new materials to support sustainable wastewater remediation treatment.

Chapter 1
Introduction

1.1 BACKGROUND AND MOTIVATION

The global population has grown significantly over the last few decades; this has led to a great increase in the demand for water supply. The water shortage situation has the potential to develop into a global water crisis. Reliance on surface water bodies alone seems to be insufficient to respond to this rising demand, while heavy extraction of ground water can lead to negative long term effects such as land subsidence. Recycle and re-use of water, therefore, has emerged into an urgent environmental and social issue. Treated wastewater is one of the promising alternatives, which has drawn great interest over the last two decades, especially in arid and semi-arid regions where water sources are limited. The wastewater is a readily available and reliable source, and its treatment and recycling would reduce the extensive amount of water extracted from the natural environment (Toze, 2006). The potential use of the treated wastewater can be varied significantly, depending upon the degree of treatment and quality for public acceptance. Advances in the treatment made it possible to use wastewater from irrigation purposes to public water supply.

Water quality is a crucial issue for reuse of treated wastewater. Generally, the treated wastewater must at least meet minimum safety standards for specific reclamation purpose. Wastewater discharged from different industries like mining, metal finishing, battery, alloy, and electroplating is often contaminated by different toxic heavy metals (Fe(II), Cu(II), Pb(II), Co(II), Cd(II), Ni(II), Zn(II). . . .) which have varieties of adverse effects on both environment and also dependent living beings. Heavy metals are elements that have more than 5 times the specific gravity than that of water (Huamain *et al.* 1999; Srivastava & Majumder, 2008). Commonly referred heavy metals are the non-alkali metals in the periodic table with an atomic number greater than 20. Unlike organic contaminants–ubiquitous nature of heavy metals, their toxicity and nephrotoxicity even in trace quantities, tendency for bioaccumulation through the food chain, and ability to attach to body cells in different forms, have made them very critical pollutants in wastewater. Toxic heavy metals of particular concern in treatment of industrial wastewaters include iron (Fe(II)), copper (Cu(II)), cadmium (Cd(II)), cobalt (Co(II)), lead (Pb(II)), nickel (Ni(II)), and zinc (Zn(II)) as because they are frequently present in wastewater. In the last decade considerable attention has been paid to remove heavy metals from different wastewater for both public and industrial use. Different treatment techniques for wastewater laden with heavy metals have been developed in recent years both to decrease the amount of wastewater produced and to improve the quality of the treated effluent. Although various treatments such as chemical precipitation, coagulation and flocculation, flotation, adsorption, ion exchange and membrane filtration can be employed to remove heavy metals from contaminated wastewater, they have their inherent advantages and limitations in application. To date, the currently available treatment technologies regarded as effective processes for reclaiming the wastewater effluent have been still experiencing a series of technical and economical challenges. This has led to extensive research of advanced technologies that can overcome such inherent limitations. Rapid urbanization, constraints of space and demand, require that new, rapid wastewater treatment technology be developed to mitigate the demand for quality water.

Quick lime (CaO) and hydrated lime ($Ca(OH)_2$) are commonly used precipitant agents due to their availability and low-cost in most countries (Mirbagheri & Hosseini, 2005; Aziz *et al.* 2008). Lime precipitation can be employed to effectively treat inorganic effluent with metal concentrations greater than 1000 mg L^{-1}. To enhance lime precipitation, fly ash was used as a seed material (Chen *et al.* 2009b). Although widely used, hydroxide precipitation has several limitations, including: (1) chemical precipitation requires a large amount of chemicals to reduce metals to an acceptable level for discharge (Jüttner *et al.* 2000), (2) hydroxide precipitation produces large volume low density sludge, which can prevent de-watering and cause disposal problems (Kongsricharoern & Polprasert, 1995), (3) some metal (zinc and lead) hydroxides are amphoteric, tending to redissolve as the pH changes beyond the optimal range and the presence of other metals create problems when mixed metals are present in wastewater, (4) pH must be strictly controlled and a sequential precipitation approach is usually needed for the treatment of complex wastewaters containing different heavy metals.

On the other hand sulphide precipitation is also an effective process using sulfidic agents such as CaS, FeS, Na_2S, and so on. The solubility of metal sulphide precipitate is significantly lower than hydroxide precipitate, and sulphide precipitates are not amphoteric in nature. Limitations of the process involve generation of toxic hydrogen sulphide (H_2S) gas when wastewater is acidic in nature (typically $pH < 4$) and the concern for sulfide toxicity. However, eliminating excess sulfide reagent prevents formation of the odor causing hydrogen sulfide. Most of the metal sulfides have lower solubility than that of the corresponding hydroxides in a pH range of 6.0 to 11.0 (Lewis, 2010; Fu & Wang, 2011). The combination of both hydroxide and sulfide precipitation for optimal metals removal is a viable option when the heavy metals load is quite high ($\geq$400 mg L^{-1}) (Charerntanyarak, 1999; Kurniawan *et al.* 2006). A common configuration of a two-stage process is hydroxide precipitation followed by sulfide precipitation. This produces the good quality effluent of the sulfide precipitation process while significantly reducing the volume of sludge generated by hydroxide precipitation and the dose requirement of precipitating reagents. Still, combined process generates colloidal precipitate that results floc and turbidity of effluent. To remove colloidal precipitates from the effluent, an additional treatment process (filtration or separation) is required. As the precipitate is colloidal in nature, it takes time to settle out of solution. As a result, the whole process becomes slow to treat large volume of mixed heavy metals from wastewater. To achieve a high quality water at high rate, another tertiary process is needed, for example, adsorption, ion exchange or membrane filtration. Among all these techniques, adsorption is more effective in terms of cost, and easier is to handle, and re-use. There are many types of adsorption possible, but magnetic adsorption has the capability to treat large volume of water rapidly.

Nanotechnology has become one of the most significant technologies of the 21st century. It encompasses a broad range of tools, technologies and applications based on a structural size between 1 and 100 nm. An unique aspect of nanotechnology is the enormously increased ratio of surface to volume presented in many nanoscale materials, leading to new possibilities in surface-based science. Due to their small size and well-organized structure, nanoscale materials offer an alteration of physico-chemical properties of the corresponding bulk material properties, for example, colour, strength, and thermal resistance and so on, providing opportunities to be exploited in many industrial facets. Nanomaterials are considered as new functional materials in environmental science and engineering with the capability of improving water treatment. Water purification is already identified as a priority area of nanotechnology applications. In the last decade, a wide range of water treatments incorporating nanosciences have been studied such as nanofiltration membranes, nanoporous/filter materials, nanocatalysts, and magnetic nanoparticles and nanosensors. Many technology developers confirm that these nanotechnologies can offer cost-effective processes for removal of aqueous pollutants (Barker *et al.* 2009). As a result of their size, nanomaterials can exhibit an array of unique novel properties which can be utilized in development of new heavy metal treatment technologies and improvement of existing ones. Some of their properties, such as higher chemical reactivity, higher surface area, self assembly, higher specificity, higher affinity, higher capacity and other properties make them an excellent candidate for rapid removal of heavy metals from complex wastewater in large scale application with small dosages.

During the conceptualization of the work, it was thought that if common lime and lime based materials could be prepared as nanomaterials, they would have large surface area per volume, higher reactivity and less dissolution time-all increasing the heavy metals removal efficiency and effectiveness. Hence, it was considered in the work that if Ca-based ($Ca(OH)_2$, CaO and CaS) nanomaterials could be synthesized, then reactivity of these materials would be increased, and the required dose would be less for such treatment of mixed heavy metal containing wastewater compared to their respective bulk materials. Also due to the high surface to volume ratio, the adsorption capacity of nanomaterials could be increased. Magnetic nanomaterials could be used for magnetic separation where the nanomaterials could act both as an adsorbent as well as a seed for settling of colloidal precipitates, resulting in rapid treatment of large volume of wastewater.

Once the nanomaterials are synthesized, an integrated hybrid process of precipitation and adsorption for large volume rapid wastewater treatment with high quality effluent is possible.

1.2 OBJECTIVE OF THIS BOOK

The aim of this book is to discuss about the recent developments in the field of Environmental Nanotechnology. Synthesis of different nanomaterials and their characterization forms one of the important aspects of such application. Implementation of different processes using nanomaterials for environmental remediation is a challenging area nowadays. Here, synthesis of Ca-based and γ-Fe_2O_3 nanomaterials and their testing in laboratory schemes of treatment of wastewater are discussed elaborately. Such research involves two major technological practices for wastewater treatment, that is, chemical precipitation and adsorption stages. The steps are as follows:

(a) Synthesis and characterization of different Ca-based nanoparticles ($Ca(OH)_2$, CaO, and CaS) to work as precipitating agents, and magnetic nanomaterials (γ-Fe_2O_3) as an adsorbent in the final stage of heavy metals adsorption and separation.
(b) Evaluation and process optimization of the nanomaterials for their heavy metal removal capability, kinetics and mechanisms in simulated mixed wastewater via batch treatment systems.
(c) Integration into a precipitation and adsorption hybrid system to test as a technology solution for wastewater treatment.

1.3 SCOPE OF THIS BOOK

The scope of this book is to discuss about the development of different (Ca-based and γ-Fe_2O_3) nanomaterials and the results of the experiments to show the efficiency of removal of heavy metals spatially iron (Fe(II)), copper (Cu(II)), cadmium (Cd(II)), cobalt (Co(II)), nickel (Ni(II)), lead (Pb(II)), and zinc (Zn(II)) which are common heavy toxic metals present in wastewater. Ca-based nanoparticles can be synthesized using chemical routes, and maghemite (γ-Fe_2O_3) nanotubes by the microwave irradiation method. In batch scale studies, the comparative performance of Ca-based nanoparticles and commercial lime is evaluated, and the process is studied for optimization using dose dependent experiments. Magnetic adsorption efficiency of heavy metals is also evaluated in batch stud to optimize the process parameters such as dose, effective pH, and the kinetics of adsorption. Furthermore, a continuous bench study of multistage treatment using the discussed Ca-based nanomaterials would show the prospects of technological adaptation.

1.4 ORGANIZATION OF THIS BOOK

This book consists of seven main chapters. **Chapter 1** is the introduction chapter. **Chapter 2** is on the introduction of nanotechnology and **Chapter 3** addresses presence of different heavy metals in wastewater and their adverse effect. **Chapter 4** discusses deferent conventional and nanotechnology based watewater treatment techniques. Different techniques for synthesis of nanomaterials and their characterization are discussed in **Chapter 5** and **Chapter 6**, respectively. **Chapter 6** also presents synthesis and characterization of Ca-based nanoparticles and γ-Fe_2O_3. Application of Ca-based nanoparticles and γ-Fe_2O_3 in heavy metals remediation from wastewater in batch scale and in continuous mode are discussed in **Chapter 7** and **Chapter 8** respectively.

Chapter 2

Introduction to nanotechnology

Nanomaterials are cornerstones of nanoscience and nanotechnology. Nanostructure science and technology is a broad and interdisciplinary area of research and development activity that has been growing explosively worldwide in the past few years. It has the potential for revolutionizing the ways in which materials and products are created and the range and nature of functionalities that can be accessed. It is already having a significant commercial impact, which will assuredly increase in the future.

2.1 WHAT IS NANOMETER SCALE?

In 1959, Physics Nobel Laureate Richard Feynman gave a talk at California Institute of Technology on the occasion of the American Physical Society meeting. The talk was entitled 'There's Plenty of Room at the Bottom'. Although Feynman could not predict it, this lecture was to become a central point in the field of *nanotechnology*, long before any thing related with the word *nano* had emerged. The term *nano* derives from the Greek word for *dwarf*. It is used as a prefix for any unit such as a second or a meter, and it means a billionth of the unit. Hence, a nanometer (nm) is a billionth of a meter, or 10^{-9} meters. To get a perspective of the scale of a nanometer, please see a length scale showing the nanometer context in Figure 2.1. The nanometerscale is commonly indicated as 1–100 nm, but nanoscience and nanotechnology often deals with objects bigger than 100 nm.

2.2 WHAT IS A NANOMATERIAL?

Nanoscale materials are defined as a set of substances where at least one dimension is less than approximately 100 nanometers. A nanometer is one millionth of a millimeter, approximately 100,000 times smaller than the diameter of a human hair. Nanomaterials are of interest because at this scale unique optical, magnetic, electrical, and other properties emerge. These emergent properties have the potential for great impacts in all areas of our life. One nanometer is approximately the length equivalent to 10 hydrogen or 5 silicon atoms aligned in a line. Small features permit more functionality in a given space, but nanotechnology is not only a simple continuation of miniaturization from micrometer scale down to nanometer scale.

Materials in the micrometer scale mostly exhibit physical properties the same as that of bulk form; however, materials in the nanometer scale may exhibit physical properties distinctively different from that of the bulk. The two main reasons for this change in behaviour are an increased relative surface area, and the dominance of quantum effects. At this scale, quantum effects are important in determining the properties and characteristics of the materials. By restricting ordered atomic arrangements in small volumes, materials start to be dominated by the atoms and molecules at the surfaces, often leading to properties that are different from their bulk forms. An increase in surface area (per unit mass) will result in a corresponding increase in chemical reactivity, making some nanomaterials are useful as catalysts to improve the efficiency, for example, fuel cells and batteries. As the size of matter is reduced to tens of nanometres or less, quantum effects can begin to play a role, and these can significantly change a material's optical, magnetic or electrical properties. For instance, bulk silver is non toxic, whereas nano silver particles are capable of killing viruses upon contact. The second exceptional property of nanomaterials is that they can be fabricated atom by atom, with a process called bottom up, and finally nanomaterials have an increased surface area to volume compared to bulk materials. This higher surface area to volume of materials allows using less material, which has higher environmental and economic benefits as well as fabricating highly miniaturised devices, which can be portable and will need less power to operate. So, the properties of nanomaterials are dependent on both size and chemical composition.

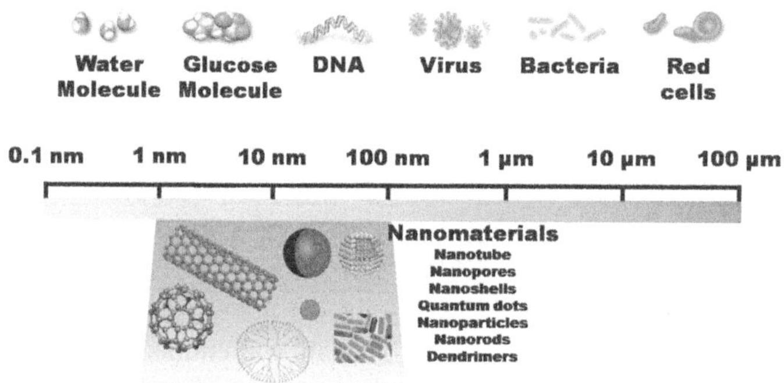

Figure 2.1 The nanoscale.

Nano-sized materials currently are used in numerous industries, for example, carbon black particles make rubber tires wear resistant; nanofibers are used for insulation and reinforcement of composites; iron oxide creates the magnetic material used in disk drives and audio–video tapes; nano-zinc oxides and titania are used as sunblocks for UV rays; and so on. Nanoscale particles and nanothin layers of materials are being used, among other things, to make products lighter, stronger and more conductive. Some of the products on the market using nanotechnology are: magnetic recording tapes, computer hard drivers, bumpers on cars, solid-state compasses, protective and glare-reducing coatings for eyeglasses and windows, automobile catalyc converters, metal-cutting tools, dental bonding agents, long lasting tennis ball, burn and wound dressing, ink and so on. Promising applications of nanotechnology in medicine and/ or biology have attracted a lot of attention and have become a fast growing field. One of the attractive applications in nanomedicine is the creation of nanoscale devices for improved therapy and diagnostics. Such nanoscale devices or nanorobots serve as vehicles for delivery of therapeutic agents, detectors or guardians against early disease and perhaps, repair of metabolic or genetic defects. For applications in medicine, the major challenge is 'miniaturization': new instruments to analyze tissues literally down to the molecular level, sensors smaller than a cell allowing to look at ongoing functions, and small machines that literally circulate within a human body pursuing pathogens and neutralizing chemical toxins.

2.3 DEFINITION OF NANOSCIENCE AND NANOTECHNOLOGY

A commom working definition of Nanoscience is the study of phenomena and manipulation of materials at atomic, molecular and macromolecular scales, where properties differ significantly from those at a larger scale.

Nanoscience is similar to materials science in that it is an integrated convergence of academic disciplines. There exist a couple of major distinctions between the two: size and biology. The objects are controlled on this size scale either in terms of manufacturing, modification or analysis, and the research includes some aspect of novelty either in terms of material studied, methods used or questions asked.

According to Royal Society of Chemistry, Nanotechnology is defined as 'Nanotechnologies are the design, characterization, production and application of structures, devices and systems by controlling shape and size at nanometre scale'.

Nanotechnology is a enabling convergent technology that cuts across all vertical industrial sectors while nanoscience is a horizontal integrating interdisciplinary science that cuts across all vertical science and engineering disciplines.

Nanotechnology is a disruptive technology with a high barrier of entry that will impact the development of enhanced materials and devices. Nanotechnology will require that a new genre of partnerships be formed among and between business, academe, and government. It will focus study and effort on potential societal implications of a new and certainly disruptive technology. Nanotechnology is predicted to signifi cantly impact the wealth and security of nations. Nanotechnology is the next industrial revolution.

Nanotechnology is considered to be, more so than ever, a technology that will have great impact on all aspects of culture and society.

2.4 HISTORY OF NANOTECHNOLOGY

The first mention of nanotechnology (not yet using that name) was in a talk given by Feynman (1959), entitled '*Theres Plenty of Room at the Bottom*'. Feynman suggested a means to develop the ability to manipulate atoms and molecules directly, by developing a set of one-tenth-scale machine tools analogous to those found in any machine shop. These small tools would be used to develop and operate a next generation of one-hundredth-scale machine tools, and so forth. As the sizes get smaller, it would be necessary to redesign some tools because the relative strength of various forces would change. Gravity would become less important, surface tension would become more important, van der Waals attraction would become important, and so on. Feynman mentioned these scaling issues during his talk. The feasibility of his proposal has never been effectively refuted.

The first use of the term 'nanotechnology' was by Taniguchi (1974) at the International Conference on Precision Engineering (ICPE). His definition referred to 'production technology to get extra high accuracy and ultra fine dimensions, that is, the preciseness and fineness on the order of 1 nm (nanometer), 10–9 m in length.' Ideas of nanotechnological strategy, which were put forward by Feynman, were developed by E. Drexler in his book 'Vehicles of creation: the arrival of the nanotechnology era' (1986).

Although Feynman and Drexler certainly popularized nanotechnology, their influence did not directly lead to the designe of nanoscale materials. Rapid progress in nanotechnology could only take place after the arrival of sophisticated instrumentation, capable of viewing and manipulating materials on the nanoscale. The 1986 Nobel Prize in Physics was awarded to Gerd Binnig and Heinrich Rohrer to honor their design of the scanning tunneling microscope (STM). They shared the Prize with Ernst Ruska, the inventor of the first electron microscope, another essential tool for the modern nanomaterials scientist. In fact, the resolution of modern electron microscopes are now high enough to provide images of individual atoms, and are often fitted with detectors that are capable of determining the chemical composition and/or oxidation state of the surface atoms.

2.5 CLASSIFICATION OF NANOSTRUCTURED MATERIALS

There are several ways to classify nanomaterials. This depends very much on the applications and philosophy which are adopted. We shall just present here two extreme classifications which can be found in the literature. In the first one, the focus will be on the composition of the material. For the purpose of this article, most current nanomaterials could be organized into four types depending on composite of nanomaterials:

(1) Carbon Based Materials
(2) Metal Based Materials
(3) Dendrimers
(4) Composites

2.5.1 Carbon based materials

These nanomaterials are composed mostly of carbon, most commonly taking the form of a hollow spheres, ellipsoids, or tubes. Spherical and ellipsoidal carbon nanomaterials are referred to as fullerenes, while cylindrical ones are called nanotubes. These particles have many potential applications, including improved films and coatings, stronger and lighter materials, and applications in electronics. This classification includes fullerenes, carbon nanotubes, graphene and the likes.

2.5.2 Metal based materials

These nanomaterials include quantum dots, noble metal nanomaterials and metal oxides, such as zinc oxide. A quantum dot is a closely packed semiconductor crystal comprised of hundreds or thousands of atoms, and whose size is on the order of a few nanometers to a few hundred nanometers. Changing the size of quantum dots changes their optical properties. For example, titanium dioxide (TiO_2) nanoparticles are extensively used in applications such as paint, sunscreen, and toothpaste.

2.5.3 Dendrimers

These nanomaterials are nanosized polymers built from branched units. The surface of a dendrimer has numerous chain ends, which can be tailored to perform specific chemical functions. This property could also be useful for catalysis. Also, because three-dimensional dendrimers contain interior cavities into which other molecules could be placed, they may be useful for drug delivery.

2.5.4 Composites

Composites combine nanoparticles with other nanoparticles or with larger, bulk-type materials. Nanoparticles, such as nanosized clays, are already being added to products ranging from auto parts to packaging materials, to enhance mechanical, thermal, barrier, and flame-retardant properties.

Another classification has the focus on the macroscopic dimensions of the material. Nanomaterials are then defined as materials containing structures having at least one dimension less than about 100 nm. A nanostructure is said to have one dimension, for example, if it has a length larger than 100 nm in one direction only. A wire or a fiber is a 1-dimensional object and a thin film a 2-dimension nanostructure. To summarize, 0D nanomaterials contain spheres or clusters which are considered as point-like particles.

1D nanomaterials contain nanofibers, wires, rods, and so on. 2D are films, plates, multilayers, or networks. 3D nanomaterials are nanophase materials consisting of equiaxed nanometer-sized grains. This is shown schematically in Figure 2.2.

Dimension	Structure	Example
0-D All dimensions (X,Y,Z) at nanometric scale	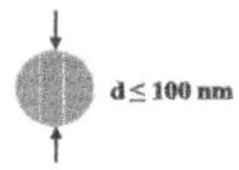	Nanoparticles
1-D Two dimensions (X,Y) at nanometric scale, one other dimension (L) is not	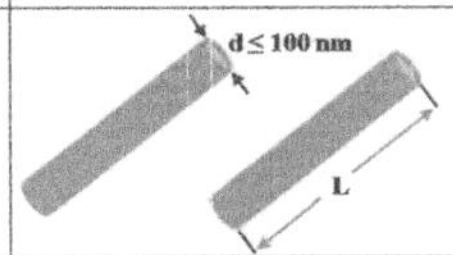	Nanotubes and Nanorods
2-D One dimension (t) at nanometric scale, two dimension (L_X, L_Y) is not	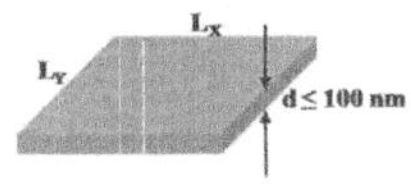	Thin nanofilms
3-D All three dimensions (L_X, L_Y, L_Z) are not at nanometric scale	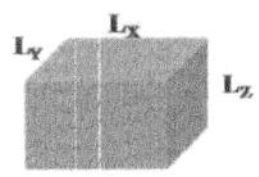	Nanocrystalline and nanocomposite materials

Figure 2.2 Classification of materials according to the dimensionality of the nanostructures.

2.6 UNIQUE PROPERTIES OF NANOMATERIALS

The unique properties of these various types of intentionally produced nanomaterials give them novel electrical, catalytic, magnetic, mechanical, thermal, or imaging features that are highly desirable for applications in commercial, medical, military, and environmental sectors. These materials may also find their way into more complex nanostructures and systems. As new uses for materials with these special properties are identified, the number of products containing such nanomaterials and their possible applications continue to grow.

Between the dimensions on an atomic scale and the normal dimensions, which characterize bulk material is the size range where condensed matter exhibits some remarkable specific properties that may be significantly different from the physical properties of bulk materials. Some such typical properties are known, but there may be a lot more to be discovered. Some known physical properties of nanomaterials are related to different origins: for example, (i) large fraction of surface atoms, (ii) large surface energy, (iii) spatial confinement, and (iv) reduced imperfections. The following are just a few examples.

Nanomaterials exhibit many unique properties, for which they are intensely being studied in a number of research fields. For example, already discussed that they have very high surface area to volume ratio compared to the bulk material. Room-temperature superplasticity in nanocrystalline metals facilitates the fabrication process for producing components with complex shapes. Nanoparticles can also enhance strength and uniformity of composite materials. Lastly, they show quantum confinement effects that form the basis in developing high technology devices. If a significant change in the atomic structure, electronic, magnetic, and optical properties of the material is observed, the chemical reactivity of the particle will also be significantly affected. The factors that contribute to size dependent change in chemical reactivity and properties of a material can be explained through the following interrelated mechanisms:

(a) *Size reduction*: the proportion of atoms at the surface or near surface regions to those before increases considerably when NPs size decreases. This causes a more reactive surface.
(b) *Change in surface free energy*: the increasing reactive surface leads to a change in surface free energy with respect to particle size, thus influencing the chemical reactivity.
(c) *Atomic structure variation*: when the size decreases, defects on and near the surface in the form of change in vacancies, bond length, and bond angle will occur.
(d) *Change in electronic structure*: as the size gets smaller and smaller, the electronic structure resembles discrete energy states of small molecules (Wigginton *et al.* 2007).

Chapter 3

Heavy metals and their presence in wastewater

3.1 HEAVY METAL POLLUTION IN THE AQUATIC ENVIRONMENT

Wastewaters encompass many types of effluents: industrial waste, sewage effluent, mining waste, and so on. Industrial wastewater can be contaminated with a wide variety of toxic chemicals, with heavy metals and organics being the most significant and potentially the most troublesome contaminants. Unlike organic toxicants, inorganic heavy metal species are non-biodegradable. 'Heavy metals' are named after those metals between atomic number 21 (Scandium) and atomic number 84 (Polonium), except for Aluminum (Al), which has atomic number 13, but it is also considered a heavy metal (Schnoor, 1996). A valid definition for the term 'heavy metals' has never been established (Duffus, 2002). Nor has the term 'trace metals', which is often used synonymously, ever been defined exactly (Kabata-Pendias, 2000). Several sources define heavy metals as elements with a density greater than 5 g cm^{-1} (Parker, 2002; Brewer & Scott, 1983; Lozet & Mathieu, 1990; Morris, 1992). Sadler *et al.* (1985) suggested that under this definition all metals in the periodic table, especially Group I and II should be considered as heavy metals. According to Huamain *et al.* (1999) and Srivastava and Majumder (2008) heavy metals are elements that have more than 5 times the specific gravity of water. Alloway (1995) specifies that only metals of industrial significance and natural occurance must be included to this definition. These are cadmium (Cd(II)), copper (Cu(II)), chromium (Cr(III)), cobalt (Co(II)), iron (Fe(II)), mercury (Hg(II)), molybdenum (Mo(II)), nickel (Ni(II)), lead (Pb(II)), selenium (Se(II)), and zinc (Zn(II)). These elements have the potential to be toxic to fauna and humans, and cause neurological disorder, mutagenesis, cancer, among other problems and diseases. They can accumulate in the food chain and pose great dangers to living organisms, including plants, animals and humans. Heavy metal contamination in the aquatic environment, including Cd(II), Cu(II), Pb(II), Fe(II), Co(II) and Zn(II), pose an eco-toxicological hazard to living organisms. Some of these metal pollutants come from fertilizers and sewage but the largest source is definitely the industrial effluent discharged from various industries, such as mining, alloy industry, metal plating industries, electroplating and battery manufacturing. The ingestion of these toxic metals beyond the permitted concentration may bring about an increased risk of serious health disorders. Table 3.1 shows the water quality standard for the heavy metals.

Table 3.1 Primary drinking water standards.

Heavy metal	EPA discharge limit, (mg L^{-1}) (USEPA, 1999)	The National Environment Regulations, (mg L^{-1}) (BIS, 1999)	Risk-based drinking water criteria, (μg L^{-1}) (WHO, 1993)
Cd(II)	2.0	0.1	3
Cu(II)	3.0	1.0	2000
Pb(II)	0.1	0.1	10
Ni(II)	3.0	1.0	20
Fe(II)	3.0	10	EPA secondary criteria: 300
Co(II)	–	–	50
Zn(II)	5	5	5000

Due to the cumulative toxicities of heavy metals, effluents from metal-processing industries need to be pre-treated or detoxified before being discharged into the aqueous environment in order meet the regulatory requirement.

3.2 SOURCES OF HEAVY METALS

The aquatic environment and it's water quality is considered as the main factor controlling the state of health and disease in both cultured and wild fishes. Pollution of the aquatic environment by inorganic and organic chemicals is a major factor posing serious threat to the survival of aquatic organisms, including those that are used as human food like fishes and amphibians.

Heavy metals may enter the aquatic environment, such as rivers and lakes, from various sources. The first source can be from the nature itself. Wet and dry fallout of atmospheric particulate matters derived from the natural source, such as the dust from the weathering of rock and soil, or from human activities, including the combustion of fossil fuels and the processing of metals, can introduce relatively a large percentage of the heavy metals in rivers and lakes. Dead and decomposing vegetation and animal matter also contribute to a small percentage of the metals in the adjacent waters. Generally, groundwater has higher dissolved mineral concentrations than surface waters. This is due to the intimate contact between the CO_2-bearing water and the rocks and soils in the ground that are rich in metal compounds and the long length of contact time leading dissolution (Dix, 1981).

The main point sources of heavy metal pollution may be attributed to the anthropogenic factors. Heavy metals exist in aqueous waste streams of many industries, such as from electroplating operations, mining industrial activities, and power-generating stations, and so on. (Sengupta, 2001). The waste streams produced from these industries have sometimes been left behind and hence polluted the surrounding soils, surface water and ground water. Santos *et al.* (2002) reported the heavy metal pollution of groundwater in the alluvial aquifers of the Guadiamar River in Spain by Aznalcollar mine tailing spill. The excessive utilization of some heavy metals contained agrochemicals in agriculture can be another source of heavy metal pollution. Wong *et al.* (2002) reported the accumulation of heavy metals in agricultural soils of Pearl River delta, south China in the past few decades, owing to the rapid urban and industrial development and increasing reliance on agrochemicals. The heavy metals contained in soils leached into water because of rain, and caused pollution of the Pearl River.

Pollution of the aquatic environment by inorganic chemicals has been considered a major threat to the aquatic organisms including fishes. The agricultural drainage water containing pesticides and fertilizers and effluents of industrial activities and runoffs in addition to sewage effluents supply the water bodies and sediment with huge quantities of inorganic anions and heavy metals (ENV, 2002). The most anthropogenic sources of metals are industrial, petroleum contamination and sewage disposal (Santos *et al.* 2005).

3.3 TOXICOLOGY OF COMMON HEAVY METALS

The list of heavy metals is quite large. Here those heavy metals, are discussed which will later be discussed in the form experiments for removal from wastewater. Needless to say, most of the heavy metals show similar properties.

3.3.1 Cadmium (Cd)

Cadmium occurs naturally in the earth's crust and is usually found in combination with zinc. Cadmium is released into rivers through weathering of rocks, some of it is released into air through forest fires and volcanoes, and the rest of it is released through anthropogenic activities, such as manufacturing. Nearly three-fourths of the quantity of cadmium is used in Ni–Cd batteries while most of the remaining one-fourth is used for pigments, coating, electroplating and as stabilizing agents for plastics (Reddy *et al.* 2008). Cadmium enters our bodies mainly through uptake of food and it tends to accumulate in the human body with ~33% in the kidney and ~14% in the liver. Itai-Itai, was once one of the four big pollution diseases in Japan, associated with cadmium poisoning (Volesky, 1990). Other adverse health effects that can be caused by this toxic metal include stomach pain, reproductive failures, psychological disorders, nephrotoxicity, and immune system and nervous system disorders. Moreover, animal studies have demonstrated an increase in lung cancer from long-term inhalation exposure to cadmium and the U.S. EPA has classified cadmium as a probable human carcinogen (USEPA, 1999).

3.3.2 Copper (Cu)

Copper is a very common metal that exists naturally in the environment and spreads across the environment through natural phenomena. Two types of copper can be formed: Cu(I) and Cu(II). As the industries advanced, copper has become one of the most widely used metals. The primary waste sources of copper discharged from industries include printed circuit boards, metal finishing industries, tannery operations, chemical manufacturing, and mining drainage (Stephenson & Blackburn, 1998). The production of copper has increased over the last decades, which contributes to the expanded copper quantities in the environment. Excess amount of copper in the body can inhibit the enzyme dihydrophil hydratase, an enzyme involved in haemopoiesis. An inherited condition called Wilson's disease causes the body to retain copper, since it is not excreted by the liver into the bile. If this disease is not treated, brain and liver damage can occur. Excessive copper in water has also been found to damage marine life. The observed effect of higher concentrations of copper to fish and other creatures is damage to gills, liver, kidneys, and the central nervous system.

3.3.3 Lead (Pb)

Lead exist in three oxidation states: Pb(0), Pb(II), and Pb(IV). Metallic lead, Pb(0) exists in nature, but its occurrence is rare. In the environment, lead is commonly found to combine with two or more other elements to form lead compounds and it usually exists as Pb(II). Lead occurs naturally in the earth's crust but the high level of lead found throughout the environment mainly comes from human activities. The use of leaded gasoline, house paint, and pesticides has already been banned in recent years; lead is still widely used in car batteries and ammunitions. The improper disposal of wastewater from these industries will then contribute to the rise in lead level in the environment. Accumulation of lead in the human body causes a rise in blood pressure, abortion, improper function in the brain and kidney as well as behavioral disruption and mental retardation of young children. Moreover, as lead can enter a fetus through the placenta of the mother, it can also cause damage to the brain and the central nervous system of unborn children (Volesky, 1990).

3.3.4 Nickel (Ni)

Nickel is widely distributed in the environment and is the twenty-fourth most abundant element in the earth's crust. The major uses of refined nickel in industry include electroplating, alloy production and fabrication, the manufacture of nickel–cadmium batteries and electronic components, and the preparation of catalysts for hydrogenation of fats and methanation. Acute nickel toxicity effects include gastrointestinal symptoms such as nausea, vomiting, abdominal discomfort and diarrhea and neurological symptoms like headache, giddiness, coughing and shortness of breath. Nickel salts affect the pulmonary and digestive tract of humans. Exposure to soluble nickel compounds also causes respiratory tract cancer if the dosage exceeds 1 μg of Ni L^{-1} (Rana, 2006). Nickel also affects the blood, liver, kidney and immune systems (Zhang *et al.* 2006c). There is some evidence of lung and nasal sinus cancers in humans due to Nickel. Metallic nickel is teratogenic and carcinogenic to mammals (Rana, 2006). Higher concentrations of nickel cause cancer of lungs, nose and bone. Dermatitis (Ni itch) is the most frequent effect of exposure to Ni, such as coins and jewellery. Acute poisoning of Ni(II) causes headache, dizziness, nausea and vomiting, chest pain, tightness of the chest, dry cough and shortness of breath, rapid respiration, cyanosis, and extreme weakness (Al-Asheh & Duvnjak, 1997; Kadirvelu *et al.* 2001; Beliles, 1979).

3.3.5 Iron (Fe)

Iron is defined as an essential micronutrient due to its involvement in the formation of hemoglobin. It is found in the earth's crust, often in combination with other chemical elements. Some iron compounds are considered to be soluble in the whole pH range, ferric and ferrous humates. The availability of iron increases with decreasing pH (Sauchelli, 1969). Iron is an essential nutrient, which is involved in the oxygen transport inside human beings. Proper functioning of the immune system relies, in part, on sufficient amounts of iron. The body stores iron very efficiently, and too much iron can be toxic. Iron overload is a condition characterized by excessive absorption of iron through diet. Much of the additional iron is deposited in the liver, damaging this vital organ and causing it to enlarge. Iron overload can lead to a vast range of symptoms like, joint pain, constant fatigue, heart problems, hair loss, abnormal liver function, impaired memory, mood swings, and depression. Some studies suggest that iron overload may cause cancer, such as colorectal cancer. Harmful levels of iron may also be deposited in other organs, including the heart and the pancreas. It can also affect joints, leading to symptoms similar to arthritis.

3.3.6 Cobalt (Co)

Cobalt exists in the principal minerals smaltite ($CoAs_2$) and caboltite (CoAsS) and it only represents 4×10^{-3} percent of the earth's crust. Cobalt commonly exists in two oxidation states, (II) and (III). In oxygenated waters the (II) state is immediately oxidized to the (III) state but since the (III) state hydrolyzes, its observed concentrations are low (Pagenkopf, 1983). Cobalt is used in the preparation of magnetic, wear-resistant, and high-strength alloys. Its compounds are used in ink, paint, and varnish industries. Cobalt is also a central component of the vitamin cobalamin, or vitamin B12 and it is only slightly toxic. The isotope ^{60}Co is commonly used in radiotherapy, sterilization of medical supplies, and medical waste. But this high-energy gamma emitting isotope can cause severe burns and death. Extended exposures increase the risk of morbidity or mortality from cancer.

3.3.7 Zinc (Zn)

Zinc is a common element occurring naturally in the environment and it is widely used by humans for domestic and industrial purposes. Zinc compounds are widely used in industry to make paint, rubber, dyes, and wood preservatives. Depending on the type of soil, some zinc compounds can percolate into the groundwater and into lakes, streams and rivers. The solubility of zinc is highly dependent on the compounds, for example zinc phosphate is insoluble and zinc nitrate is very soluble (Sauchelli, 1969). Low levels of zinc, Zn(II), are essential for maintaining good health. If zinc levels are too low, it will results in failure to eat, skin lesions, sexual immaturity, depression, and can affect taste (Alloway, 1995).

3.3.8 Arsenic (As)

Arsenic (As) is now a toxic pollutant found in different parts of world. Arsenic has many applications: (a) in bronzing, hardening and improving the sphericity of shot, wood preservation, pyrotechnics, varieties of semiconductor devices solar cells, light-emiting diodes, lasers, and integrated circuits); and (b) as pesticides (for Toxic Substances & ATSDR, 2000). Arsenic has been reported to vary in: (a) blood, from 1.5 to 2.5 $\mu g\,L^{-1}$; (b) hair, from 0.25 to 0.88 $\mu g\,L^{-1}$; and (c) urine the average concentrations are between 20 and 50 $\mu g\,L^{-1}$ (Apostoli, 1999; Yamato, 1988). In the environment arsenic is usually found combined with other elements in inorganic and organic forms. Inorganic arsenic is known to be more poisonous than organic one. Arsenic trioxide (As_2O_3) is the most common inorganic arsenical in air, while arsenates (AsO_4^{3-}) or arsenites (AsO_2) occur in water, soil, or food. Arsenic may be also necessary ultra-trace element for red algae, chickens, rats, goats, and pigs and its deficiency inhibits growth. Arsenic concentration is high in marine food. In fishes arsenic ranged between 5 to 100 $\mu g\,g^{-1}$ and reach up to 100 to 250 $\mu g\,g^{-1}$ in species at the top of the food chain (Michel, 1993). It has been documented that in 1955 Arsenic was accidentally mixed into the Morinag's Powdered Milk 'MF' (Japan) due to which 600 new born babies died and 624 were afflicted by severe mental retardation, developmental difficulties, and braindamage-related paralysis (Ui, 1992).

In humans arsenic toxicity occurs due to ingestion of As containing powders or solutions in accidents, suicide, homicide, or consumption of contaminated food or drinking water. Arsenic has been reported to be associated with hypertension and serious impacts on the cardiovascular system, and even hepatic damage at high doses (Lee *et al.* 2003; Yoshida *et al.* 2004). It has a suppressive effect on spermatogenesis and gonadotrophin and testosterone release in rats (Sarkar *et al.* 2003). There is correlation between arsenic exposure and diabetes mellitus (type II) (Walton *et al.* 2004). Besides, inorganic arsenic ingestion arsenic leads to various dermal effects like: hyperkeratosis, hyperpigmentation and hypopigmentation; periorbital swelling; the occurrence of spontaneous abortion and damage of the nervous system (if high doses are taken in).

3.3.9 Mercury (Hg)

Mercury is used as a component of barometers, thermometers, dental products (amalgam), electrical equipment and in control devices, as well as in fungicides. It is also used in some of the gold industry. Mercurous chloride (calomel) is one of the oldest known pharmaceuticals and is continuously used for its antiseptic properties. It prevents seeds from fungus contamination and is used to amalgamate other metals. Thimerosol is antiseptic containing 49.5% ethyl mercury that has been used for years as a preservative in many infant vaccines and in flu vaccines (James *et al.* 2005). Extensive use of wood from forests results in soil erosion, which contributes to the accumulation of inorganic and alkylated mercury components in the aquatic ecosystem (Webb, 2004). This could result in increased methylmercury (MeHg) concentrations in fish that can be of concern in exposed groups of the population, because methylmercury is almost completely absorbed into the bloodstream Nordberg *et al.* (1985). Any source of environmental mercury represents a potential risk for human MeHg poisoning, because the methylation of inorganic mercury to MeHg in waterways results ultimately in its accumulation in the sea food chain, which represents the most prevalent source for human consumption (Sanfeliu *et al.* 2003; Shanker *et al.* 2003).

The 'Minamata disease' is the most known incident of organic mercury poisoning (Weiss, 1995). The pollution with methylmercury has shown the importance of biologically mediated transformation reactions resulting in organometallic compounds (Chang, 1977; Annau & Cuomo, 1988; Bellama *et al.* 1988; Sanfeliu *et al.* 2003).

Recently the safety of thimerosal, an ethyl mercury-containing preservative used in vaccines, has been questioned due to exposure of infants during immunization. Mercurials have been reported to cause apoptosis in cultured neurons; however, the signalling pathways resulting in cell death have not been well characterized. *Cytochrome c* was shown to leak from the mitochondria upon organic mercury exposure (thiomersal), followed by caspase 9 cleavage, caspase 3 activation, deleterious effects on the cytoarchitecture and initiation of mitochondrialmediated apoptosis (Humphrey *et al.* 2005). Because mercury has a high affinity for thiol (sulfhydryl (–SH)) groups, the thiol-containing antioxidant, glutathione (GSH), provides the major intracellular defence against mercury-induced neurotoxicity (James *et al.* 2005). Astrocytes are targeted by MeHg toxicity and they increase neuronal resistance (Shanker *et al.* 2003; Morken *et al.* 2005). It was also shown that skeletal muscle is an important deposit of MeHg (Gonzalez *et al.* 2005).

3.3.10 Chromium (Cr)

Chromium is the most abundant of all the heavy metals on the earth that can also be found in wastewater. It is discharged from the wastewater of the steel manufacturing industry, paints industry, leather tannery products, dye and textile industry, paper industry, electroplating, and chrome plated products (Malkoc *et al.* 2006; Coey, 2009). Chromium exists in two forms: trivalent Cr(III) and hexavalent Cr(VI). Its abundance is harmful to the marine life, vegetation and plants, in addition to humans due to its toxicity (Yavuz *et al.* 2006b). Cr(VI) is also a strong oxidizing agent, as well as a potential carcinogen. Other potential harmful effects on humans due to long term exposure to the metal include liver damage, kidney circulatory damage, nerve tissue damage, and dermatitis. In general, chromium results in very grave harmful effects on human health (Yavuz *et al.* 2006b). Cr(III) is less toxic than Cr(VI) as it only affects plants, not humans. Thus, the total concentration of Chromium in the form of both Cr(VI), Cr(III), in addition to other forms, is regulated to a threshold of 2 mg L^{-1}. The reduction of chromium concentration is crucial in order to meet this threshold. Hence, the U.S. Environmental Protection Agency (EPA) regulates a threshold of maximum allowable concentration of the metal to be discharged into water bodies without treatment to be 0.05 mg L^{-1}. The US EPA has also set the maximum contamination level for chromium to be 0.1 mg L^{-1} in drinking water.

Chapter 4

Treatment techniques of heavy metals in wastewater

4.1 CONVENTIONAL TREATMENT OF METAL–LADEN WASTEWATER

There are many methods that are currently being used for treating heavy metals rich wastewater (Figure 4.1). Some of the commonly used methods are discussed below.

4.1.1 Chemical precipitation

Chemical precipitation is effective and by far the most widely used process in the industry (Fu & Wang, 2011) because it is relatively simple and inexpensive to operate. In precipitation processes, chemicals react with heavy metal ions to form insoluble precipitates. The forming precipitates can be separated from the water by sedimentation or filtration. The treated water is then decanted and appropriately discharged or reused. The conventional chemical precipitation processes include hydroxide precipitation and sulfide precipitation.

4.1.1.1 Hydroxide precipitation

The most widely used chemical precipitation technique is hydroxide precipitation due to its relative simplicity as a process, low cost and ease of pH control (Huisman *et al.* 2006; Lewis, 2010; Matlock *et al.* 2002; Pérez *et al.* 2010; Meunier *et al.* 2006; González-Muñoz *et al.* 2006). The solubilities of the various metal hydroxides are minimized in the pH range of 8.0–11.0. The metal hydroxides can be removed by flocculation and sedimentation. A variety of hydroxides has been used to precipitate metals from wastewater, based on the low cost and ease of handling; lime is the preferred choice of base used in hydroxide precipitation at industrial settings (Baltpurvins *et al.* 1997). The conceptual mechanism of heavy metal removal by chemical precipitation is presented in Equation 4.1.1

$$M^{2+} + 2(OH)^{-} = M(OH)_2 \downarrow \tag{4.1.1}$$

where M^{2+} and OH^{-} represent the dissolved metal ions and the precipitant, respectively, while $M(OH)_2$ is the insoluble metal hydroxide.

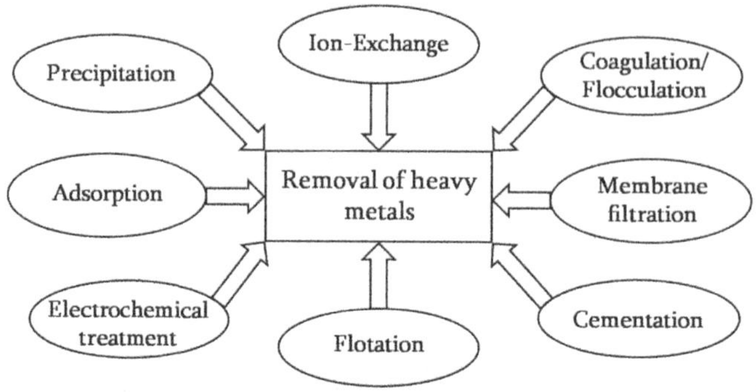

Figure 4.1 Some conventional treatment methods for the removal of heavy metals from solution.

Hydroxide precipitation process using $Ca(OH)_2$ and NaOH in removing Cu(II) and Cr(VI) ions from wastewater was evaluated by Mirbagheri and Hosseini (2005). In the hydroxide precipitation process, the addition of coagulants such as alum, iron salts, and organic polymers can enhance the removal of heavy metals from wastewater. Charerntanyarak (1999) employed chemical coagulation and precipitation by lime to treat synthetic wastewater consisting of Zn(II), Cd(II), Mn(II) and Mg(II) at concentration of 450, 150, 1085, and 3154 mg L^{-1}, respectively. In spite of their varying initial concentrations, almost complete removal from synthetic wastewater was achieved for all the metals at pH 11.0, complying with the effluent limit of the Thai Pollution Control Department for Zn(II) and Mn(II) of less than 5 mg L^{-1}. Moreover, if coagulant was added, the residual concentration of heavy metals could be decreased further.

Overall, pH adjustment to the basic conditions (pH 11.0) is the major parameter that significantly improves heavy metal removal by chemical precipitation. Due to its availability in most countries, lime or calcium hydroxide is the most commonly employed precipitant agent. Lime precipitation can be employed to effectively treat inorganic effluent with a metal concentration of exceeding 1000 mg L^{-1}. Other advantages of using lime precipitation include the simplicity of the process, inexpensive equipment requirement, and convenient and safe operations, making it a popular method for metal removal from contaminated wastewater. It was reported that values of pH 9.5 or greater are necessary for effective precipitation of Zn(II), Fe(II), Cu(II), Mn(II) and Mg(II) as hydroxide with lime treatment. But the same procedure may be unsatisfactory for Co(II), Ni(II), and Cd(II) and the use of sulfide for Co(II), Ni(II), and Cd(II) precipitation appears to be necessary (Charerntanyarak, 1999; Tavlarides *et al.* 1987).

Although widely used, hydroxide precipitation has several limitations. The limitations are as follows:

(a) Hydroxide precipitation generates large volumes of relatively low density sludge, which can prevent dewatering and disposal problems.
(b) Some metal (Zn(II) and Pb(II)) hydroxides are amphoteric, tend to re-dissolve as the pH changes beyond the optimal range and the mixed metals create a problem using hydroxide precipitation since the ideal pH for one metal may put another metal back into solution.
(c) The pH must be strictly controlled and a fractional precipitation approach is usually needed for the treatment of complex wastewaters containing different heavy metals.
(d) Hydroxide precipitation requires a large amount of chemicals to reduce metals to an acceptable level for discharge.
(e) Hydroxide precipitation producing large volume low density sludge, which can prevent de-watering and cause disposal problems (Kongsricharoern & Polprasert, 1995).
(f) Chemical precipitation requires a large amount of chemicals to reduce metals to an acceptable level for discharge (Jüttner *et al.* 2000).

4.1.1.2 Sulfide precipitation

Sulfide precipitation is another promising precipitation process to treat toxic heavy metals from wastewater. The primary advantage of sulfide precipitation over hydroxide precipitation is the solubility of metal sulfide is much lower compared to metal hydroxides (Figure 4.2 and Table 4.1) and also generates precipitates that are not amphoteric in nature.

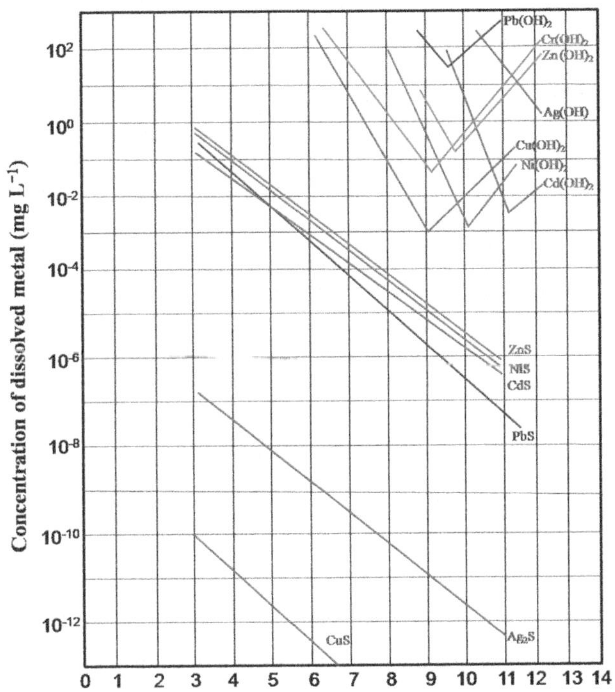

Figure 4.2 Metal solubility vs pH in hydroxide and sulfide compound.

Table 4.1 Solubility product of selected metals as hydroxide and sulfide at 25°C.

Metal	As Hydroxide	As Sulfide
Cd(II)	2.3×10^{-5}	6.7×10^{-10}
Cu(II)	2.2×10^{-2}	5.8×10^{-18}
Pb(II)	2.1	3.8×10^{-9}
Ni(II)	6.9×10^{-3}	6.9×10^{-8}
Fe(II)	8.9×10^{-1}	3.4×10^{-5}
Co(II)	2.2×10^{-1}	1.0×10^{-8}
Zn(II)	1.1	2.3×10^{-7}

Hence, the sulfide precipitation process can achieve a high degree of metal removal over a broader pH range compared to hydroxide precipitation. Metal sulfide precipitate/sludge also exhibits better thickening and dewatering characteristics than the corresponding metal hydroxide precipitate/sludges.

Metal sulfide precipitation can be expressed as:

$$H_2S \longleftrightarrow HS^- + H^+ \quad (4.1.2)$$

$$HS^- \longleftrightarrow S^{2-} + H^+ \quad (4.1.3)$$

$$M^{2+} + S^{2-} \longleftrightarrow MS(s) \downarrow \quad (4.1.4)$$

$$M^{2+} + HS^- \longleftrightarrow MS(s) \downarrow + H^+ \quad (4.1.5)$$

Bhattacharyya *et al.* (1979) used Na_2S as the sulfide form to separate the metal cations Cd(II), Zn(II), Cu(II) and Pb(II) and oxyanions of Ar and Se precipitate from complex wastewaters at concentrations between 0.02–0.3 mM. Jong and Parry (2003) treated mildly acidic metal (Cu(II), Zn(II), Ni(II), Fe(II), and Mg(II)), arsenic and sulfate waters using a bench scale, Upflow Anaerobic Packed Bed (UAPB) reactor filled with silica sand. The initial metal concentrations were approximately 10 mg L^{-1} (0.1–0.8 mM). The selective separation and recovery of copper, cadmium, zinc, and nickel from a polymetallic solution with sulfide precipitation using thioacetamide have been investigated by Gharabaghi *et al.* (2012).

However, sulfide precipitation is not used as widely as it could be because the dosing of sulfide is seen as being difficult to control (due to the very low solubility of the metal sulfides and thus the sensitivity of the process to the dose) and because of concerns about the toxicity and corrosiveness of excess sulfide (Veeken *et al.* 2003; Peters & Bennett, 1989). As it is well known both, heavy metal ions are often present in acidic conditions and sulfide precipitants in acidic conditions can result in the evolution of toxic H_2S fumes. It is essential that this precipitation process be performed in a neutral or basic medium. Moreover, metal sulfide precipitation tends to form colloidal precipitates that cause some separation problems in either settling or filtration processes.

4.1.2 Ion exchange

Ion exchange is a process in which the metal ions are removed from the aqueous phase by the exchange of cations or anions between the exchange medium and the wastewater (Clifford, 1999; Smara *et al.* 2007). The materials used for making ion exchange media are either natural or synthetic organic materials or inorganic polymeric materials. These resins can be regenerated for reuse after their exchange capacity is exhausted.

The uptake of heavy metal ions by ion-exchange resins is affected by certain parameters such as pH, temperature, initial metal concentration and contact time (Gode & Pehlivan, 2006). Many researchers have demonstrated that zeolites exhibit good cation-exchange capacities for heavy metal ions under different experimental conditions (Motsi *et al.* 2009; Ostroski *et al.* 2009; Taffarel & Rubio, 2009). Even though this is an efficient method for wastewater treatment, it is not that widely used because of its high cost. Another disadvantage of this method is that during the ion exchange process, these resins can absorb and store all the toxic chemicals which need to be disposed.

4.1.3 Coagulation/flocculation

Coagulation and flocculation followed by sedimentation and filtration is also employed for removal of heavy metal from wastewater (Akbal & Camcı, 2010; Chang *et al.* 2009; Heredia & Martín, 2009; Hankins *et al.* 2006). Coagulation is the destabilization of colloids by neutralizing the forces that keep them apart. Many coagulants are commonly used in the conventional wastewater treatment processes including such coagulants aluminium, ferrous sulfate and ferric chloride, resulting in the effective removal of wastewater particulates and impurities by charge neutralization of particles and by enmeshment of the impurities on the formed amorphous metal hydroxide precipitates.

El Samrani *et al.* (2008) investigated the removal of heavy metal by coagulation of combined sewer overflow with two commercial coagulants, a ferric chloride solution and a polyaluminium chloride (PAC). Flocculants of Konjac–graft–poly (acrylamide)–co–sodium xanthate (Duan *et al.* 2010) and polyampholyte chitosan derivatives–N–carboxyethylated chitosans (Bratskaya *et al.* 2009) have also been used to remove heavy metals. Generally, coagulation/flocculation can not treat the heavy metal wastewater completely (Chang & Wang, 2007). Therefore, coagulation/flocculation must be followed by other treatment techniques. Plattes *et al.* (2007) employed precipitation, coagulation and flocculation processes using ferric chloride to remove tungsten from industrial wastewater.

4.1.4 Membrane filtration

Membrane filtration has received considerable attention for the treatment of inorganic effluent, since it is capable of removing not only suspended solids and organic compounds, but also inorganic contaminants such as heavy metals. Depending on the size of the particle that can be retained, various types of membrane filtration such as ultrafiltration, nanofiltration and reverse osmosis can be employed for heavy metal removal.

Ultrafiltration (UF) utilizes a permeable membrane to separate heavy metals, macromolecules and suspended solids from inorganic solution on the basis of the pore size (5–20 nm) and molecular weight of the separating compounds (1000–100,000 Da). To obtain high removal efficiency of metal ions, the micellar enhanced ultrafiltration (MEUF) and polymer enhanced ultrafiltration (PEUF) was proposed. Samper *et al.* (2009) used MEUF to remove Cd(II), Cu(II), Ni(II), Pb(II), and Zn(II) from synthetic water using two anionic surfactants: SDS and linear alkylbenzene sulfonate (LAS) in a laboratory-scale membrane system.

Nanofiltration (NF) is considered as an intermediate process between UF and reverse osmosis (RO). NF is a promising technology for the rejection of heavy metal ions such as nickel (Murthy & Chaudhari, 2008), chromium (Muthukrishnan & Guha, 2008), copper (Cséfalvay *et al.* 2009; Ahmad & Ooi, 2010) and arsenic (Nguyen *et al.* 2009; Figoli *et al.* 2010) from wastewater.

The reverse osmosis (RO) process uses a semi-permeable membrane, allowing the fluid that is being purified to pass through the membrane by the use of pressure, while rejecting the contaminants. RO is one of the techniques that is able to remove a wide range of dissolved species from water. It accounts for more than 20% of the world's desalination capacity (Shahalam *et al.* 2002). Cu(II) and Ni(II) ions were successfully removed by the RO process and the rejection efficiency of the two ions was as high as up to 99.5% by using Na_2EDTA (Mohsen-Nia *et al.* 2007).

4.1.5 Cementation

Cementation is another type of another precipitation method implying an electrochemical mechanism in which a metal having a higher oxidation potential passes into solution (for example, oxidation of metallic iron, Fe(0) to ferrous Fe(II) to replace a metal having a lower oxidation potential). Copper is most frequently separated by cementation along with noble metals such as Ag(II), Au(II) and Pb(II). As(III), Cd(II), Ga(II), Pb(II), Sb(II), and Sn(II) can be recovered in this manner.

4.1.6 Flotation

Flotation has nowadays found extensive use in wastewater treatment (Lundh *et al.* 2000; Tassel *et al.* 1997; Tessele *et al.* 1998; Yuan *et al.* 2008; Polat & Erdogan, 2007; Gopalratnam *et al.* 1992). Flotation has been employed to separate heavy metal from a liquid phase using bubble attachment, a technique that originated in mineral processing. Dissolved air flotation (DAF), ion flotation and precipitation flotation are the main flotation processes for the removal of metal ions from solution.

DAF had been widely studied to remove heavy metal in 1990s (Waters, 1990; Tassel *et al.* 1997; Tessele *et al.* 1998). Ion flotation has been shown as a promising method for the removal of heavy metal ions from wastewaters. Yuan *et al.* (2008) investigated the potential of ion flotation to remove cadmium, lead and copper from dilute aqueous solution with a plant-derived biosurfactant tea saponin. The maximum removal of Pb(II), Cu(II), and Cd(II) can reach 89.95%, 81.13% and 71.17%, respectively, when the ratio of collector to metal was 3:1. The precipitate flotation process is another alternative flotation method, based on the formation of precipitate and subsequent removal by attachment to air bubbles.

4.1.7 Electrochemical treatment

Electrochemical methods involve the plating-out of metal ions on a cathode surface and can recover metals in the elemental metal state. Electrochemical wastewater technologies involve a relatively large capital investment and the expensive electricity supply, so they have not been widely applied. There are several different types of electrochemical treatment process available such as electrocoagulation (EC), electroflotation (EF), electrochemical precipitation (EP) and so on.

Electrocoagulation (EC) involves the generation of coagulants in-situ by dissolving electrically, either aluminum or iron ions from aluminum or iron electrodes (Chen, 2004). Heidmann and Calmano (2008) studied the performance of an EC system with aluminium electrodes for removing Zn(II), Cu(II), Ni(II), Ag(I/II), and $Cr_2O_7^{2-}$.

Electroflotation (EF) is a solid/liquid separation process that floats pollutants to the surface of a water body using tiny bubbles of hydrogen and oxygen gases, generated from water electrolysis. EF has wide range applications in heavy metals removal from industrial wastewater. Belkacem *et al.* (2008) investigated the clarification of wastewater using the EF technique with aluminum electrodes. The application of the optimized parameters on the separation of some heavy metal ions such as iron, nickel, copper, zinc, lead and cadmium was studied. Their study demonstrated that the metal removal rate reached as high as 99%.

Electrochemical precipitation (EP) has been used to maximize the removal of heavy metal from contaminated wastewater; electrical potential has been utilized to modify the conventional chemical precipitation.

4.1.8 Adsorption

Adsorption is one of the more effective methods for removing heavy metals from industrial wastewaters. Adsorption is a surface phenomenon involving binding molecules or particles in a solution onto a surface. Adsorption is mainly of two types: physical adsorption and chemical adsorption. The physical adsorption process does not involve the sharing or transfer of electrons. The interactions are fully reversible enabling desorption to occur at the same temperature and is not site specific. However, chemical sorption involves chemical-like bonding onto the surface of adsorbents and is typically an irreversible process. Chemisorption is site specific, that is, the chemisorbed molecules are fixed at specific sites.

Activated carbon (AC) adsorbents are widely used in the removal of heavy metal contaminants. Its usefulness derives mainly from its large micropore and mesopore volumes and the resulting high surface area. A large number of researchers have studied the use of AC for heavy metal removal (Jusoh *et al.* 2007; Kang *et al.* 2008; Ku & Peters, 1987). Researchers investigated industrial by-products such as lignin (Betancur *et al.* 2009; Reyes *et al.* 2009), diatomite (Sheng *et al.* 2009), clino pyrrhotite (Lu & Liu, 2006), lignite (Mohan & Chander, 2006), aragonite shells (Köhler *et al.* 2007), natural zeolites (Apiratikul & Pavasant, 2008), clay (Al-Jlil & Alsewailem, 2009), kaolinite (Gu & Evans, 2008) and peat (Liu *et al.* 2008b), and so on.

Many varieties of low-cost adsorbents have been developed and tested to remove heavy metal ions. However, the adsorption efficiency depends on the type of adsorbents and its adsorption capacity in single- and multi-component systems (Lin *et al.* 2000; Peters & Ku, 1987). Carbon nanotubes (CNTs) discovered by Iijima, (1991) in 1991, have been widely studied for their excellent properties and applications. As relatively new adsorbents, CNTs have been proven to possess great potential for removing heavy metal ions such as lead (Wang *et al.* 2007; Kabbashi *et al.* 2009), cadmium (Kuo & Lin, 2009), chromium (Pillay *et al.* 2009), copper (Li *et al.* 2010c), and nickel (Kandah & Meunier, 2007) from wastewater.

4.1.8.1 Adsorption equilibria

The equilibrium adsorption capacity, q_e (mg g^{-1}) of heavy metal ions aqueous was calculated using the mass balance equation:

$$q_e = \frac{(C_0 - C_e) \times V}{m} \qquad (4.1.6)$$

where C_0 and C_e (mg L^{-1}) are initial and equilibrium concentrations of metal ions, respectively; V (L) is the sample volume, m (g) is the mass of adsorbent.

4.1.8.2 Langmuir isotherm model

The Langmuir Adsorption Isotherm is used for monolayer adsorption onto a surface containing a finite number of identical sites, and assumes uniform energies of adsorption on the surface, in addition to no transmigration of the adsorbate in the plane of the surface (Langmuir, 1916). The Langmuir isotherm model determines the maximum capacity of the adsorbent from complete monolayer coverage of the adsorbent surface. The Langmuir's isotherm non-linear equation is

$$q_e = \frac{q_m \times b \times C_e}{(1 + b \times C_e)} \qquad (4.1.7)$$

The linearized Langmuir isotherm is given as:

$$\frac{C_e}{q_e} = \frac{1}{q_m \times b} + \frac{C_e}{q_m} \qquad (4.1.8)$$

where q_e is the equilibrium adsorption capacity of adsorbent in mg (metal ions) g^{-1} (adsorbent), C_e is the equilibrium concentration of metal ions in mg L^{-1}, q_m is the maximum amount of metal ions adsorbed in mg (metal ions) g^{-1} (adsorbent), and b is the constant that refers to the bonding energy of adsorption in L mg^{-1}. They can be calculated from the $1/q_m$ vs $1/C_e$ figure, where q_m is 1/intercept and b is equal to the intercept $\times$ 1/slope.

4.1.8.3 Freundlich isotherm model

The Freundlich Adsroption Isotherm may be used for heterogenous surface energies of adsorption. The isotherm portrays that the ratio of the quantity of the solute adsorbed on a known mass of adsorbent to the concentration of the solute in the solution differs at different concentrations (Freundlich, 1906). Langmuirs energy term, b, varies here as a function of q_e due to variations in the heat of adsorption. The Freundlich equation is as follows:

$$q_e = K_f C_e^{\frac{1}{n}} \tag{4.1.9}$$

The linearized Freundlich isotherm is given as:

$$\log q_e = \log K_f + \frac{1}{n} \log C_e \tag{4.1.10}$$

where q_e is the equilibrium adsorption capacity of the adsorbent in mg (metal ions) g^{-1} (adsorbent), C_e is the equilibrium concentration of heavy metal ions in mg L^{-1}, K_f is the constant related to the adsorption capacity of the adsorbent in mg L^{-1}, and n is the constant related to the adsorption intensity. The constants n and K_f are determined from the slope and intercept of the linear plot lnq_e versus lnC_e.

4.1.8.4 Desorption study

Desorption of absorbance from interfaces depends essentially on the conditions under which they have been adsorbed. Therefore, suitable desorption agents should be selected to release the adsorbate from magnetic particles. The extent of desorption was calculated from the following expression (Vieira & Beppu, 2006):

$$Desorption(\%) = \frac{Desorbed\ amount\ of\ metal\ ions}{Adsorbed\ amount\ of\ metal\ ions} \times 100 \tag{4.1.11}$$

4.2 DEVELOPMENT OF NANOTECHNOLOGY IN WATER AND WASTEWATER TREATMENT

Water on Earth is the most precious substance for all life and finite resource and it is continuously recycled in the water cycle. In recent time water has become a scarce resource, and for many countries supplies already fall short of demand. Due to climate change and growth of population in the urban areas, water has become even scarcer. Moreover, physical, chemical, or biological properties of water have been altered due to continuous addition of contaminants such as organic/inorganic materials, pathogens, heavy metals or other toxins from different sources, making it unsafe for the ecosystem and drinking, can be termed as wastewater. Furthermore, water quality standards are become more stringent due to emerging contaminants and impose new scrutiny to the existing water treatment and distribution systems which are widely established in developed countries. As a result of it, there is a growing need for clean and affordable water especially for drinking purposes and access to it is considered as one of the most important humanitarian goals and it remains as a major challenge for the 21st century. Use of unconventional water sources and reuse of wastewater have become a new form of water supplies to mitigate this challenge. Due to limitation of conventional water and wastewater treatment technologies, adequate quality water supply to meet human and environmental needs now reach their limit. This situation is demanding for technologies that produce clean and affordable water after treatment and do not cause any detrimental effect to human beings or the environment. Advancement in the area of nanotechnology creates new opportunity to develop technologies that might help alleviate water problems by solving the technical challenges.

Nanotechnology is the act of manipulating materials at very tiny scales (generally regarded as nanoscale) – essentially at the atomic and molecular size levels. When materials have one or more of their dimensions below 100 nanometers, the normal rules of physics and chemistry often no longer been apply. As a result, many materials start to display unique and sometimes, surprising properties. Their strength, ability to conduct electricity and rate of reactivity increases dramatically. For example, solids such as gold turn into liquids at room temperature, silver shows increased antimicrobial properties, inert materials like platinum and gold become catalysts, and stable materials like aluminum become combustible. These newly discovered properties of nano-scale materials have opened up exciting fields of study and applications in areas that can improve the quality of human life in the fields of water and health. Use of nanotechnology in water and wastewater treatment, not only shows its capacity to overcome different challenges associated with conventional technology but also explore novel treatment technology to treat and reuse of wastewater in respect of economic feasibility and water quality standard.

In this section of the chapter, an overview of recent advances in nanotechnology for water and wastewater treatment is discussed here. The application of nanomaterials in wastewater is critically discussed in perspective of their functional processes. Moreover, the challenges associated with the application of nanotechnology towards full-scale wastewater treatment also are discussed here.

4.2.1 Recent application of nanotechnology in water and wastewater treatment

Since water treatment by using nanoparticles has potential for application, however, its usage cost should be managed in ways to survive existing competition in the market (Crane & Scott, 2012). Due to higher surface to volume ratio they impart size dependent properties such as high reactivity, higher adsorption capacity, and higher dissolution activity can be used in wastewater treatment. In addition to that some unique property like superparamagnetism, semiconducting, and quantum confinement effect also added advantages in treatment processes. There are various recent advances on different nanomaterials (nanostructured catalytic membranes, nanosorbents, nanocatalysts, bioactive nanoparticles, biomimetic membrane and molecularly imprinted polymers (MIPs)) for removing toxic metal ions, disease causing microbes, organic and inorganic solutes from water (Figure 4.3).

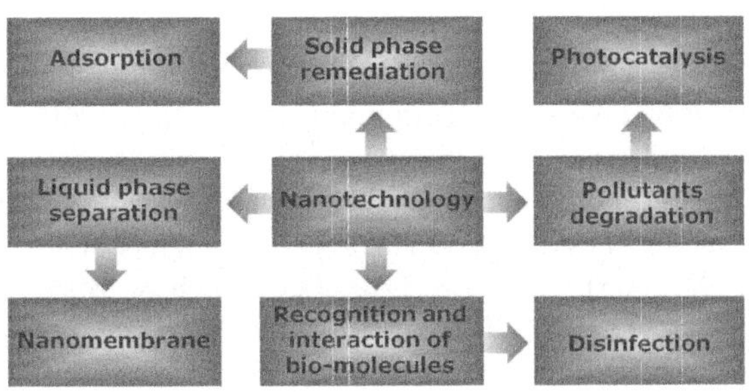

Figure 4.3 Application of nanotechnology in wastewater treatment.

4.2.2 Adsorption

Sorbents are widely used in tertiary treatment and purification of wastewater and water to remove organic and inorganic contaminants. Efficiency of conventional sorbents which are used in adsorption processes such as activated carbon, ion-exchange resins and so on. suffer due to lack of high available surface area or active sites, lack of selectivity and specificity, and lack of adsorption kinetics. The use of nanomaterials (Nanosorbent) may have advantages to overcome challenges of conventional materials (Sorbents) due to the much larger surface area of nanoparticles, higher specificity and selectivity, and tunable pore size and surface chemistry on a mass basis. In addition, the unique structure and electronic properties of some nanoparticles can make them especially powerful adsorbents. Commercialized nanosorbents are very few mainly from the U.S. and Asia but research is on going on in large numbers targeting various specific contaminants in water.

4.2.2.1 Carbon nanotubes in waste water treatment

Discovery of carbon nanotubes (CNTs) eventually revolutionize the future nanotechnologies area. CNTs as reported by Iijima (1991) and Bethune *et al.* (1993), are seamless cylinder-shaped macromolecules with a radius as small as a few nanometers, and up to several micrometers in length. The walls of these tubes are constructed of a hexagonal lattice of carbon atoms and capped by fullerene-like structures. CNTs, owing to their tunable physical, chemical, electrical and structural properties, inspire innovative technologies to address the water shortage and water pollution problems. CNTs embedded nanotechnologies have found water-treatment applications in many fields, such as sorbents, catalyst, filters or membranes. In aqueous phase, CNTs get aggregated due to hydrophobic nature and help in forming spaces and grooves for adsorption of organic molecules. This property has attracted a lot of attention as very powerful adsorbents for a wide variety of organic compounds from water for example, dioxin (Long & Yang, 2001), polynuclear aromatic hydrocarbons (PAHs) (Yang *et al.* 2006c, b; Gotovac *et al.* 2006), DDT and its metabolites (Zhou *et al.* 2006c), PBDEs (Wang *et al.* 2006a), chlorobenzenes and chlorophenols (Peng *et al.* 2003; Cai *et al.* 2005), trihalomethanes (Cai *et al.* 2005; Lu *et al.* 2006, 2005), bisphenol A and nonylphenol (Cai *et al.* 2003a), phthalate esters (Cai *et al.* 2003b), dyes (Fugetsu *et al.* 2004), pesticides (thiamethoxam, imidacloprid and acetamiprid) (Zhou *et al.* 2006a) and herbicides such as sulfuron derivatives (Zhou *et al.* 2006b, 2007), atrazine (Zhou *et al.* 2006d) and dicamba (Biesaga & Pyrzynska, 2006). Cross-linked nanoporous polymers that have been copolymerized with functionalized CNTs have been demonstrated to have a very high sorption capacity for a variety of organic compounds such as p-nitrophenol and trichloroethylene (Salipira *et al.* 2007). It was found that purification (removal of amorphous carbon) of the CNTs improved the adsorption (Gotovac *et al.* 2006). The available adsorption space was found to be the cylindrical external surface; neither the inner cavity nor the inter-wall space of multi-walled CNT contributed to adsorption (Yang & Xing, 2007). Unlike the case with fullerenes, no adsorption–desorption hysteresis was observed, indicating reversible adsorption (Yang & Xing, 2007).

For the removal of metals and other inorganic ions, mainly nanosized metal oxides (Pacheco *et al.* 2006; Deliyanni *et al.* 2007) but also natural nanosized clays (Yuan & Wu, 2007) have been investigated. Also, oxidized and hydroxylated CNTs are good adsorbers for metals. This has been found for various metals such as Cu (Liang *et al.* 2005), Ni (Lu & Liu, 2006; Chen & Wang, 2006), Cd (Li *et al.* 2003; Liang *et al.* 2004) and Pb (Li *et al.* 2006, 2002). Adsorption of organometallic compounds on pristine multi-walled CNTs was found to be stronger than for carbon black (Muñoz *et al.* 2005). Many studies have focused on the removal of heavy metal ions by modified carbon nanotubes, including cadmium (Vuković *et al.* 2010), nickel and strontium (Chen *et al.* 2009a), lead (Vuković *et al.* 2011), chromium (Atieh, 2011), uranium (Schierz & Zänker, 2009), copper (Li *et al.* 2010c), and copper, zinc, cadmium, and nickel (Salam *et al.* 2011) ions from aqueous solutions.

4.2.2.2 Metal nanomaterials in wastewater treatment

Nanoparticles formed by metal or metal oxides are another type of inorganic nanomaterials, which are used broadly to remove heavy metal ions in wastewater treatment. Nanosized metals or metal oxides include nanosized silver nanoparticles (Fabrega *et al.* 2011), ferric oxides (Feng *et al.* 2012), titanium oxides

(Luo *et al.* 2010), magnesium oxides (Gao *et al.* 2008), copper oxides (Goswami *et al.* 2012), cerium oxides (Cao *et al.* 2010), and so on, all these provide high surface area and specific affinity. Besides, common metal oxides possess little environmental impact due to low solubility and thus, no secondary pollution; they have been adopted as sorbents to remove heavy metals. Hristovski *et al.* (2007) worked on the feasibility of arsenate removal by aggregated metal oxide nanoparticle media in packed bed columns. Through batch experiments conducted with 16 commercial nanopowders in four water matrices, TiO_2, Fe_2O_3, ZrO_2, and NiO nanopowders are selected on the basis of fitted Freundlich adsorption isotherm parameters, to exhibit the highest arsenate removal in all water matrices. Singh *et al.* (2011a) reported on the removal of various toxic metal ions, such as Co^{2+}, Ni^{2+}, Cu^{2+}, Cd^{2+}, Pb^{2+}, Hg^{2+}, and As^{3+} from wastewater by porous ZnO nano-assemblies. It was reported that Hg^{2+}, Pb^{2+}, and As^{3+} have stronger attraction towards ZnO nano-assemblies due to their high electronegativity and, hence, that they exhibit better removal efficiency (63.5% Hg^{2+}, 100% Pb^{2+}, and 100% As^{3+}).

Recently, there have been several reports on magnetic oxides, especially Fe_3O_4, being used as nano-adsorbents for the removal of various toxic metal ions from wastewater, such as Ni^{2+}, Cr^{3+}, Cu^{2+}, Cd^{2+}, Co^{2+}, Hg^{2+}, Pb^{2+}, and As^{3+} (Hakami *et al.* 2012; Zhao *et al.* 2010; Wang *et al.* 2012; Liu *et al.* 2008a; Feng *et al.* 2012; Singh *et al.* 2011b). Shen *et al.* (2009) for example, have observed that the adsorption efficiency of Ni^{2+}, Cu^{2+}, Cd^{2+}, and Cr^{6+} ions by Fe_3O_4 nanoparticles is strongly dependent on pH, temperature, the amount of the adsorbent and the incubation time.

4.2.3 Photocatalysis in wastewater treatment

Photo-catalyst activated by UV lights can oxidize organic pollutants into non-toxic products, such as carbon dioxide, water and can also disinfect certain bacteria. Photo-Catalysis is defined as 'acceleration of photoreaction in presence of a catalyst'. A catalyst does not change in itself or get consumed in the chemical reaction.

In catalyzed photolysis, light is absorbed by an adsorbed substrate. In photo generated catalysis the photo catalytic activity (PCA) depends on the ability of the catalyst to create electron-hole pairs, which generate free radicals (hydroxyl radicals: OH) able to undergo secondary reactions (Figure 4.4). It is an useful pretreatment for hazardous and non-biodegradable contaminants to enhance their biodegradability. Photocatalysis can also be used as a polishing step to treat recalcitrant organic compounds. The major barrier for its wide application is the slow kinetics due to limited light fluence and photocatalytic activity. Si, TiO_2, ZnO, WO_3, CdS, ZnS, $SrTiO_3$, SnO_2, WSe_2, Fe_2O_3, and so on. can be used as photocatalysts and Table 4.2 shows band gap energies of those catalysts.

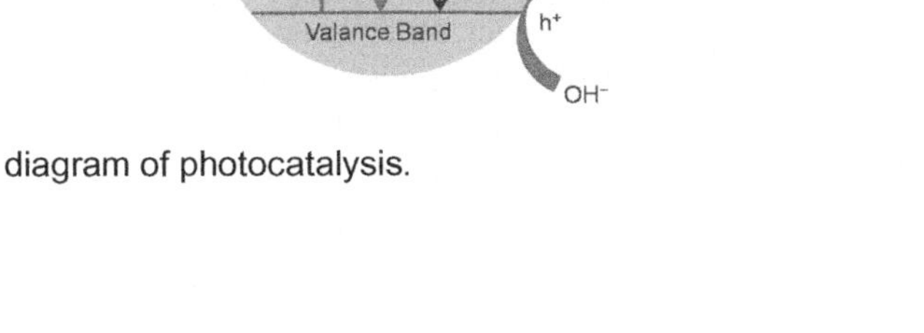

Figure 4.4 Schematic diagram of photocatalysis.

Table 4.2 Bandgap energy of various photocatalysts (Bhatkhande *et al.* 2002).

Photocatalyst	Bandgap energy (eV)	Photocatalyst	Bandgap energy (eV)
Si	1.1	ZnO	3.2
TiO_2 (rutile)	3.0	TiO_2 (anatase)	3.2
WO_3	2.7	CdS	2.4
ZnS	3.7	$SrTiO_3$	3.4
SnO_2	3.5	WSe_2	1.2
Fe_2O_3	2.2	α–Fe_2O_3	3.1

4.2.3.1 ZnO nanomaterials as photocatalyst

Zinc oxide has received a great deal of attention in relation to the photocatalytic degradation of organic contaminants. It has been reported that different morphologies of ZnO exhibit different degrees of photocatalytic activity (Gupta *et al.* 2011; Li & Wang, 2009; Ma *et al.* 2011; Kim & Huh, 2011; Li *et al.* 2010b; Xu *et al.* 2009; Zhai *et al.* 2012; Zheng *et al.* 2009; Singh *et al.* 2013). Gupta *et al.* (2011) have studied the photodegradation of MB over different-shaped ZnO nanostructures, observing that the photocatalytic activity is dependent on defect concentration. Ma *et al.* (2011) and Zhai *et al.* (2012) have reported on the photocatalytic activity of ZnO nano-rod arrays on arbitrary substrates and ZnO nano-disks in decomposing methyl orange (MO). Zheng *et al.* (2009) investigated the photocatalytic activity of octahedron and rod-like porous ZnO architectures for the decomposition of MO in water under UV irradiation. Further, they noted the high catalytic efficiency of porous ZnO octahedron calcined at 500°C, rather than 700°C due to their large surface area.

There have been various reports on the enhancement of photocatalytic performance by doping impurities (Ag, Cu, I) (Yıldırım *et al.* 2013; Mohan *et al.* 2012; Barka-Bouaifel *et al.* 2011). Mohan *et al.* (2012) demonstrated the photocatalytic activity of pure and Cu-doped ZnO nanorods for the degradation of resazurin dye. They observed a significant enhancement of photocatalytic activity upon the doping of Cu into ZnO nano-rods. However, there are also reports on the suppression of the catalytic efficacy of ZnO nanostructures upon doping with transition metal ions (Barick *et al.* 2010; Qiu *et al.* 2008; Ullah & Dutta, 2008). Barick *et al.* (2010) have observed a decrease in the photocatalytic activity of mesoporous ZnO nano-assemblies after doping with transition metal ions (Mn, Co and Ni) under UV light. Qiu *et al.* (2008) also found that Co^{2+} doping markedly suppressed the photodegradation of RhB under UV irradiation. It is proposed that the substitutions of transition metal ions in a ZnO lattice may act as trapping or recombination centres for electrons and holes and, hence, substantially decreases the photodegradation efficiency. Ullah & Dutta (2008) have reported the lower photodegradation efficiency of MB over Mn-doped ZnO as compared to pristine ZnO due to the faster recombination of electron-hole pairs following a change of the absorption characteristics caused by Mn^{2+} doping.

4.2.3.2 TiO_2 nanomaterials as photocatalyst

Titanium dioxide is another highly favourable material for heterogeneous photocatalytic processes due to its high photoactivity, non-toxic nature, large band-gap and stability. There have been numerous reports on the photoabsorption and photocatalytic properties of TiO_2 under UV light (Xie *et al.* 2010; Pang & Abdullah, 2013). Xie *et al.* (2010) have studied the photocatalytic activity of TiO_2 at three different temperatures (120, 160 and 200°C) and found the highest activity at 160°C. They also doped TiO_2 with Si and observed that Si doping does not improve the photocatalytic activity of TiO_2. However, it has been reported that the photocatalytic activity of TiO_2 can be enhanced, either by doping with transition metal ions (Fe, Bi, Ag and V) and rare-earth metal ions (Nd, Gd), or by the surface modification of the crystalline structure, as they could significantly influence charge carrier recombination rates and interfacial electron-transfer rates (Bzdon *et al.* 2012; Li *et al.* 2010a). Pang and Abdullah (2013) have demonstrated that Fe-doped TiO_2 nanotubes are an efficient candidate for the purification of real textile wastewater containing a mixture of organic dyes (which included reactive, vat and disperse dyes) as compared to TiO_2 powder and TiO_2 nanotubes.

The enhancement of the visible light photocatalytic activity of TiO_2 with the doping of non-metallic elements, such as carbon (C), nitrogen (N), boron (B) and sulphur (S) has also been reported in the literature (Yang *et al.* 2012; Pozan & Kambur, 2013; Bagwasi *et al.* 2013; Lv *et al.* 2011; Radoičić *et al.* 2013; Dong *et al.* 2011; Wu *et al.* 2009; Liu *et al.* 2012; Yu *et al.* 2011; Nolan *et al.* 2012; Diker *et al.* 2011; Cheng *et al.* 2012; Senthilnathan & Philip, 2010; Wang *et al.* 2010; Li *et al.* 2012; Zheng *et al.* 2011; Ohno *et al.* 2004; Yang *et al.* 2009; Ju *et al.* 2013). Cheng *et al.* (2012) have reported on the complete photodegradation of MB after 120 minutes by carbon self-doped TiO_2 sheets under visible light. Ju *et al.* (2013) have also observed a higher catalytic activity of 3% for N- and S-doped TiO_2 nanoparticles in the degradation of MO at pH 4 under sunshine irradiation.

4.2.3.3 Other nanomaterials as photocatalyst

To date, semiconductors such as WO_3, ZnS, CdS, Fe_2O_3, SnO_2 have been found to have sufficient band-gap energies for promoting photocatalytic activities. Zhang *et al.* (2007a) have evaluated photocatalytic activity of ZnS by decomposing methyl orange and reported that these sensitized ZnS powders with a proper molar ratio showed higher photocatalytic activity than TiO_2(P25) under visible light ($\lambda > 420$ nm) irradiation. It has been reported that coupling of CdS (band gap 2.4 eV) with SnO_2 (band gap 3.5 eV) could produce hydrogen under visible irradiation (Gurunathan *et al.* 1997). The use of iron oxides as photocatalysts has been given particular attention due to their absorptivity in the visible region, abundance, and low cost. Faust *et al.* (1989) examined the suitability α-Fe_2O_3 as a photocatalyst by studying the kinetics and mechanisms of the photocatalytic oxidation of sulfur dioxide in aqueous colloidal suspensions of 3–25 nm α-Fe_2O_3.

4.2.4 Nanomembrane in wastewater treatment

Traditional water treatment methods include physical separation techniques for particle removal; biological and chemical treatments to remove suspended solids, organic matter and dissolved pollutants or toxins; and evaporative techniques and other physical and mechanical methods. Membrane separation replaces or supplements these techniques by the use of selectively permeable barriers, with pores sized to permit the passage of water molecules, but small enough to retain a wide range of particulate and dissolved compounds, depending on their nature. Membrane filtration processes are classified according to the membrane pore sizes, which dictate the size of the particles they are able to retain (Table 4.3).

However, the effectiveness of many membranes can be improved; for example by enhancing the flux or by reducing membrane fouling. Membrane characteristics that can be improved by the use of nanoparticles are: antifouling and antimicrobial properties; increased selectivity; increased flux (through increased hydrophobicity); or poresize control by nanogels (shrinking/growing by change of pH or temperature). The application of nanoparticles in membranes can thus decrease the energy demand, the use of chemicals for membrane cleaning, and cost.

Table 4.3 Type of membrane and characteristic.

Filtration class	Particle capture size	Typical contaminants removed	Typical operating pressure ranges
Microfiltration (MF)	0.1–10 μm	Suspended solids, bacteria, protozoa	0.1–2 bar
Ultrafiltration (UF)	ca. 0.003–0.1 μm	Colloids, proteins, polysaccharides, most bacteria, viruses (partially)	1–5 bar (cross-flow), 0.2–0.3 bar (dead-end and submerged)
Nanofiltration (NF)	ca. 0.001 μm	Viruses, natural organic matter, multivalent ions (including hardness in water)	5–20 bar
Reverse osmosis (RO)	ca. 0.0001 μm	Almost all impurities, including monovalent ions	10–100 bar

To address these challenges, membranes can be functionalized with nanoparticles to improve membrane performance. This method is rather new, so it is not surprising that several different terms are being used to describe this technology; nano-activated, nano-enhanced, nanoparticle-enhanced, nanoparticle-based, and nano-functionalised membranes, and so on. To avoid any ambiguity, the terminology should be standardized. The term 'nano-enhanced membrane (NEM)' since 'enhancement' best describes the function of the nanoparticles in the membrane. Recent studies on membrane nanotechnology have focused on synthesis of nanomaterials embedded polymeric or inorganic membranes for multifunction purpose. Nanomaterials are generally used for multifunctional applications include metal oxide nanoparticles (e.g., Al_2O_3, TiO_2, and zeolite), antimicrobial nanoparticles (for example, nano-Ag and CNTs), and photocatalytic nanomaterials (e.g., bi-metallic nanoparticles, TiO_2, ZnO).

Use of metal oxide nanoparticles in NEM is to resist membrane from fouling and increase water permeability due to hydrophilic nature of metal oxide nanoparticles. Photocatalytic membrane has drawn attention in recent years due its simultaneous application of photocatalytic degradation of organic matter and continuous discharge of treated water without loss of nanophotacatalyst. The photocatalytic membrane can be prepared using different nanomaterials and synthesis methods. The TiO_2/Al_2O_3 composite membranes (Bosc *et al.* 2005; Zhang *et al.* 2006a,b; Choi *et al.* 2006, 2007), TiO_2 supported on polymer and metallic membranes (Kim *et al.* 2003b; Bellobono *et al.* 2005b,a) and doted polymer membranes containing TiO_2 particles entrapped membranes Artale *et al.* (2000), Kleine *et al.* (2002), Molinari *et al.* (2004) were studied for wastewater treatment. Also, the possible TiO_2 organic and inorganic ceramic membranes have been investigated Kwak *et al.* (2001); Kim *et al.* (2003b); Yang and Wang (2006), Wu *et al.* (2005a), Wu and Ritchie (2008). In most studies, however, photocatalytic membranes may encounter various technical problems such as membrane structure deterioration, low photocatalytic activity and loss of deposited TiO_2 layer over time. Metallic/bi-metallic catalyst nanoparticles such as nano zero-valent iron (nZVI) and noble metals supported on nZVI have been incorporated into polymeric membranes for reductive degradation of contaminants, particularly chlorinated compounds (Wu *et al.* 2005a; Wu & Ritchie, 2008). nZVI serves as the electron donor and the noble metals catalyze the reaction.

4.2.5 Antimicrobial activity

Microbial contamination of water poses a major threat to public health. With the emergence of microorganisms resistant to multiple antimicrobial agents (Barnes *et al.* 2006) there is increased demand for improved disinfection methods. Current advances in the field of nanobiotechnology, the ability to prepare nanomaterials of specific size and shape, are likely to lead to the development of antibacterial agents for application in wastewater. Reducing the size of the materials is an efficient and reliable tool for improving biocompatibility (Mirkin & Taton, 2000; Kim *et al.* 2007). However, little is known about how the biological activity of certain materials changes as the size of the constituting particles decreases to nanoscale dimensions. The functional activities of nanoparticles are influenced largely by the particle size. Antibacterial activity is related to compounds that locally kill bacteria or slow down their growth, without being in general toxic to surrounding tissue.

4.2.5.1 Metal nanomaterials

4.2.5.1.1 Ag Nanoparticles The unique physical and chemical properties of silver nanoparticles (AgNPs) make them excellent candidates for a number of day-to-day activities, and also the antimicrobial and anti-inflammatory properties make them excellent candidates for many purposes in the medical field. However, there are studies and reports that suggest that nanosilver can allegedly cause adverse effects on humans as well as the environment (Panyala *et al.* 2008). AgNPs are attractive because they are non-toxic to the human body at low concentrations and have broad-spectrum antibacterial actions (Baker *et al.* 2005). In fact, it is well known that Ag^+ ions and Ag-based compounds are toxic to microorganisms, possessing strong biocidal effects on at least 12 species of bacteria including multi-resistant bacteria like Methicillin-resistant *Staphylococcus aureus* (MRSA), as well as multidrug-resistant *Pseudomonas aeruginosa*, ampicillin-resistant *E. coli O157: H7* and erythromycin-resistant *S. pyogenes* (Lara *et al.* 2010; Shahverdi *et al.* 2007; Sondi & Salopek-Sondi, 2004) suggesting that AgNPs are effective broadspectrum (Rai *et al.* 2009) biocides against a variety of drug-resistant bacteria, which makes them a potential candidate for use in pharmaceutical products and medical devices that may help to prevent the transmission of drug-resistant pathogens in different clinical environments (Lara *et al.* 2010; Yamanaka *et al.* 2005). Recently, Mecking and co-workers demonstrated that hybrids of silver nanoparticles with amphiphilic hyperbranched macromolecules exhibited effective antimicrobial surface coating agent properties (Aymonier *et al.* 2002). AgNPs show powerful bactericidal properties even in far lower concentration (Mishra & Kumar, 2009). Moreover it is seen that, AgNPs show no significant cytotoxicity against human-derived monocyte cell lines, suggesting their use as antimicrobial additives in the process of fabrication of ambulatory and nonambulatory medical devices (Martinez-Gutierrez *et al.* 2010).

4.2.5.1.2 Au Nanoparticles Gold nanoparticlcs (AuNPs) present a higher stability when in contact with biological fluids (Peracchia *et al.* 1997). Spherical AuNPs with a variety of surface modifications are not inherently toxic to human cells, despite being taken up into them (Connor *et al.* 2005). Considering the advantageous properties, gold nanoparticles are being used to deliver protein based drugs like ampicillin for antimicrobial activities (Chen *et al.* 2006; Chamundeeswari *et al.* 2010). Another study showed that, cefaclor reduced AuNPs have potent antimicrobial activity against both Gram-positive (*Staphylococcus aurous*) and Gram-negative (*Escherichia coli*) bacteria as compared to cefaclor or AuNPs alone, The action of these novel particles is through the combined action of cefaclor inhibiting the synthesis of the peptidoglycan layer and gold nanoparticles generating 'holes' in bacterial cell walls thereby increasing the permeability of the cell wall, resulting in the leakage of cell contents and eventually cell death (Rai *et al.* 2010).

4.2.5.2 Metal oxide nanomaterials

4.2.5.2.1 CuO nanomaterials CuO nanoparticles (CuO NPs) were effective in killing a range of bacterial pathogens involved in hospital-acquired infections. But a high concentration of CuO NPs is required to achieve a bactericidal effect (Ren *et al.* 2009). It has been suggested that the reduced amount of negatively charged peptidoglycans makes Gram-negative bacteria such as Pseudomonas aeruginosa and Proteus spp. less susceptible to such positively charged antimicrobials. However, in the time-kill experiments the Gram-negative strains showed a greater susceptibility to CuO NPs combined nano Ag. Studies have been conducted to assess the potential of CuO NPs embedded in a range of polymer materials. A lower contact-killing ability was observed in comparison with release killing ability against MRSA strains. This suggests that a release of ions into the local environment is required for optimal antimicrobial activity (Ren *et al.* 2009; Cioffi *et al.* 2005). CuO NPs shows antimicrobial activity against Bacillus subtilis, methicillin resistant Streptococcus aureus, Klebsiella pneumoniae, Salmonella paratyphi and Shigella strains (Ren *et al.* 2009; Mahapatra *et al.* 2008).

4.2.5.2.2 MgO nanomaterials Magnesium oxide (MgO) prepared through an aerogel procedure (AP–MgO) yields square and polyhedral shaped nanoparticles with diameters varying slightly around 4 nm, arranged in an extensive porous structure with considerable pore volume (Klabunde *et al.* 1996). An interesting property of AP–MgO nanoparticles is their ability to adsorb and retain for a long time (in the order of months) significant amounts of elemental chlorine and bromine (Huang *et al.* 2005). The AP–MgO/X2 nanoparticles exhibited biocidal activity against certain vegetative Gram-positive bacteria, Gram-negative bacteria and the spores (Richards *et al.* 2000). AP–MgO nanoparticles are found to possess many properties that are desirable for a potent disinfectant (Koper *et al.* 2002). Because of their high surface area and enhanced surface reactivity, the nanocrystals adsorb and carry a high load of active halogens. Their extremely small size allows many particles to cover the bacteria cells to a high extent and bring halogen in an active form in high concentration in proximity to the cell (Richards *et al.* 2000). Standard bacteriological tests have shown excellent activity against *E.coli* and *Bacillus megaterium* and a good activity against spores of *Bacillus subtilis* (Koper *et al.* 2002). The bioactivity of AP-MgO/X2 nanoparticles is due to the positive charge they have in water suspension, opposite to those of the bacteria and spore cells, which enhances the total bactericidal effect.

4.2.5.2.3 TiO_2 nanomaterials TiO_2 particles catalyze the killing of bacteria on illumination by near-UV light. The generation of active free hydroxyl radicals (OH) by photoexcited TiO_2 particles is probably responsible for the antibacterial activity (Wei *et al.* 1994; Pham *et al.* 1995; Ireland *et al.* 1993). The antimicrobial effect of TiO_2 photocatalyst on *Escherichia coli* in water and its photocatalytic activity against fungi and bacteria has been demonstrated (Matsunaga *et al.* 1985, 1988; Kim *et al.* 2003a; Chawengkijwanich & Hayata, 2008).

4.2.5.2.4 ZnO nanomaterials Among the various metal oxides studied for their antibacterial activity, zinc oxide nanoparticles have been found to be highly toxic. Moreover, their stability under harsh processing conditions and relatively low toxicity combined with the potent antimicrobial properties favours their application as antimicrobials (Stoimenov *et al.* 2002). Many studies have shown that some NPs made of metal oxides, such as ZnO NP, have selective toxicity to bacteria and only exhibit minimal effect on human cells, which recommend their prospective uses in agricultural and food industries (Brayner *et al.* 2006; Thill *et al.* 2006; Reddy *et al.* 2007; Zhang *et al.* 2007b). The antimicrobial activity of zinc oxide nanoparticles have been studied against the food related bacteria *Bacillus subtilis*, *Escherichia coli* and *Pseudomonas fluorescens* (Jiang *et al.* 2009). ZnO NP could potentially be used as an effective antibacterial agent to protect agricultural and food safety from foodborne pathogens, especially *E. coli O157: H7* (Zhang *et al.* 2007b). ZnO NPs possess antimicrobial activities against *Listeria monocytogenes*, *Salmonella enteritidis* and *E. coli O157: H7* in culture media (Jiang *et al.* 2009). There are also other studies confirming the strong antimicrobial activity of ZnO nanoparticles wherein the nanoparticles could completely lyse the food-borne bacteria *Salmonella typhimurium* and *Staphylococcus aureus* (Liu *et al.* 2009). In another study, ZnO nanoparticles (12 nm) inhibited the growth of *E. coli* by disintegrating the cell membrane and increasing the membrane permeability (Jin *et al.* 2009).

4.2.5.2.5 Al_2O_3 nanomaterials Aluminum oxide NPs have wide-range applications in industrial and personal care products. The growth-inhibitory effect of alumina NPs over a wide concentration range (10–1000 $\mu g/mL$) on *Escherichia coli* have been studied (Sadiq *et al.* 2009). The antimicrobial property of these metal oxides is attributed to the generation of reactive oxygen species (ROS) which causes disruption of cell wall and subsequent cell death (Ruparelia *et al.* 2008). But alumina NPs may act as free radical scavengers. These NPs are able to rescue cells from oxidative stress-induced cell death in a manner that appears to be dependent upon the structure of the particle but independent of its size within the range of 61000 nm (Mohammad *et al.* 2008).

Chapter 5
Synthesis techniques of nanomaterials

5.1 INTRODUCTION

The synthesis of nanomaterial can be well accomplished by two broad approaches (Figure 5.1). These two approaches are 'Top–down' and 'Bottom–up'.

'Top–down' approach involves starting with a large material and etching, milling or machining a nanostructure from it by removing material. This can be done by using techniques such as precision engineering and lithography, and has been developed and utilized by the semiconductor industry for more than 30 years. Such methods offer reliability and device complexity, but they generally require higher energy. Precise engineering is a Top-down approach which is widely used in the micro-electronics industry (Dowsett, 1991), including the production of the semiconductor wafers (i.e., Si) used as substrates for computer chips and print the patterns on the ceramic wafers. Another Top-down approach method is lithography (Lan *et al.* 2007; Wu & Kumar, 2007). It is a technique involving the creation of a surface pattern by exposure to light, ions or electrons sources, and then subsequent etching and/or deposition of material onto that surface to produce desired devices (Lan *et al.* 2007). Electron- and ion-based lithography methods are capable of patterning features with the accuracy of 10 nm or less (Chou *et al.* 1997).

'Bottom-up' approach is a nanomaterial synthesis method involving the manufacturing of structures atom-by-atom. Such manufacturing is governed by certain physical rules. The approach towards achieving this include chemical synthesis, self-assembly, and positional assembly (Society and of Engineering, 2004). Chemical synthesis is a method of producing particles which can then be used either directly in production of the bulk form materials, or manufacturing more ordered and advanced materials through chemical reactions. Self-assembly is a technique involving atoms re-arranged into ordered nanoscale structures between the materials (Bogue, 2008). Positional assembly is a technique whereby atoms or molecules are deliberately manipulated and positioned (Smeitink *et al.* 2000), involving the usage of delicate nanoscale tools to build devices at the molecular level.

Synthesis of nanomaterial is most commonly done based on two methods that is,

- Chemical synthesis method
- Physical synthesis method

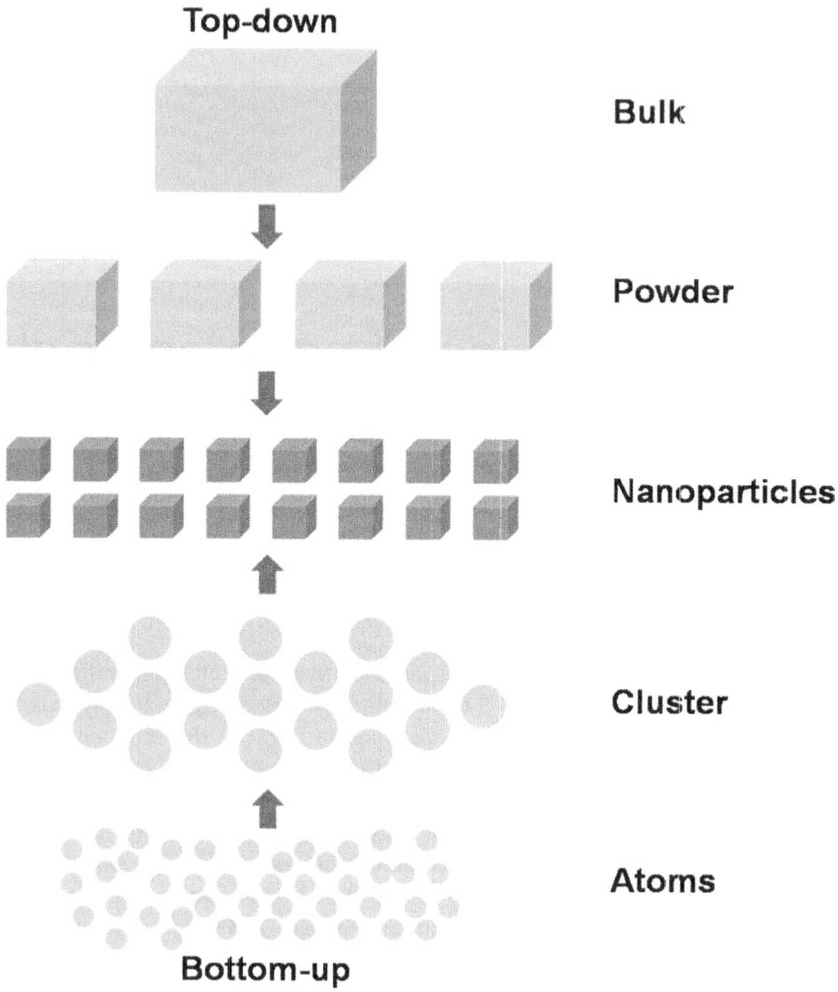

Figure 5.1 Synthesis route of nanomaterials.

5.2 CHEMICAL SYNTHESIS METHOD OF NANOMATERIALS

Chemical synthesis method of nanomaterials contains different techniques which are discussed bellow.

5.2.1 Chemical precipitation technique

Chemical synthesis has been the predominant method to developing new materials with novel properties for technological applications. It is comparatively easy to design materials of varied morphology and composition with a better control through a chemical synthesis. Since properties for a nanomaterial are very sensitive to shape, size, and composition of the material, synthesizing nanomaterials through chemical route is a preferred route. The potential of large scale synthesis of nanomaterials by chemical routes, in a cost-effective manner, is also one of the reasons behind its acceptance. $Ca(OH)_2$ nanomaterials are prepared by using chemical precipitation route (Roy & Bhattacharya, 2010).

Selection of proper reactants is an important part in any chemical synthesis process. For this purpose, extensive knowledge on chemical reactivities of the reagents, and the reaction steps that reactants are supposed to undergo, is required. The morphology and the composition of a nanomaterial can be controlled in a better way if each reaction step is known and controlled. The chemical reaction can start simply by mixing the reactants in a beaker or in a round-bottom flask. The concentration of reactants, reaction time, and order of addition of reactants to the solution, temperature, pH, viscosity, and surface tension of solution are the parameters to control. When reaction products become supersaturated, spontaneous nucleation occurs. Subsequently, it goes through the growth mechanism. Nanomaterials, with different morphology, can be prepared during this stage. The major difficulty in the chemical precipitation method is to eliminate the contamination, particularly due to the by-products generated in the chemical reactions. The optimization procedure is lengthy. Many experiments involving different parameters need to be carried out in order to get the desired result. For any chemical method, other working conditions such as stirring speed, vibration, exposure to light, cleanliness of glassware, and so on., can significantly affect the quality of the nanomaterial formed. Hence, synthesis of nanomaterials of desired morphology and composition through chemical methods is often considered to be a method involving trial and error (Nalwa, 2000).

5.2.1.1 Microwave irradiation technique

Since 1986, when Gedye and Giguere (Giguere *et al.* 1987) published their first articles in Tetrahedron Letters on microwave assisted syntheses in household microwave ovens, there has been a steadily growing interest in this research field (Mingos, 1994). Microwave heating involves the transfer of electromagnetic energy to thermal energy, and is an energy conversion phenomenon rather than that of heat transfer. Microwave irradiation is electromagnetic irradiation in the frequency range of 0.3 to 300 GHz. In general, domestic microwaves as well as the microwaves dedicated to chemical synthesis operate at a frequency of ~2.45 GHz (corresponding a wavelength of 12.24 cm). Microwave irradiation has been successfully used for organic and polymer synthesis in the past several decades (Adam, 2003; Kappe, 2002, 2004). It was found that in many cases, reaction times can be reduced in contrast to classical synthetic approaches, and the formation of undesired byproducts can be suppressed (Iannelli *et al.* 2005; Wiesbrock *et al.* 2004). Microwave-enhanced chemistry is advancing from the generation of advanced reaction conditions, for example, the pressure mediated possibility to carry out reactions above the usual boiling points of solvents, and the efficient heating of materials by selective heat absorbance. The latter process is based on the ability of suitable materials to absorb microwave energy and convert it into heat. Irradiation of the materials with microwaves results in the alignment of dipoles, ions and electrons in the applied electric field. When the applied field oscillates, the dipoles or ions attempt to align with the alternating electric field. During this process, energy is released in the form of heat due to dielectric losses and molecular friction. The amount of heat generated by this process is directly dependent on the ability of the material to align itself with the oscillating frequency of the field. Conductive materials experience a high absorption of microwave irradiation. The power absorbed in a unit volume of the material is the result of interactions between the electromagnetic field and the material, and can be expressed as,

$$P = 2\pi f \varepsilon' E^2 = \sigma E^2 \tag{5.2.1}$$

where P is the absorbed power per unit volume, f is the frequency, ε' is the complex permittivity of the material, E is the electric field intensity, and σ is the conductance. This power is related to the effectiveness of heat transfer. It decreases to 1/e of the original value at the penetration depth,

$$d_p = c\varepsilon_0/2\pi f\varepsilon' = 1/(\pi f \mu\sigma)^{1/2} \tag{5.2.2}$$

where d_p is the penetration depth, ε_0 is the dielectric constant of free space, and μ is the permeability of the material. The penetration depth is inversely proportional to the solution conductivity. Thus, effective heating is assumed for materials with higher conductivity and dimensions that are not exceeding the penetration depth (Cheng, 1989; Gaponov-Grekhov & Granatstein, 1994; Chan & Reader, 2000). Conductors have a small penetration depth and therefore reflect microwaves at their surface, whereas insulators appear to be transparent to the microwave irradiation. It offers a clean, cheap and convenient method of heating often resulting in higher yields and shorter reaction times (Figure 5.2).

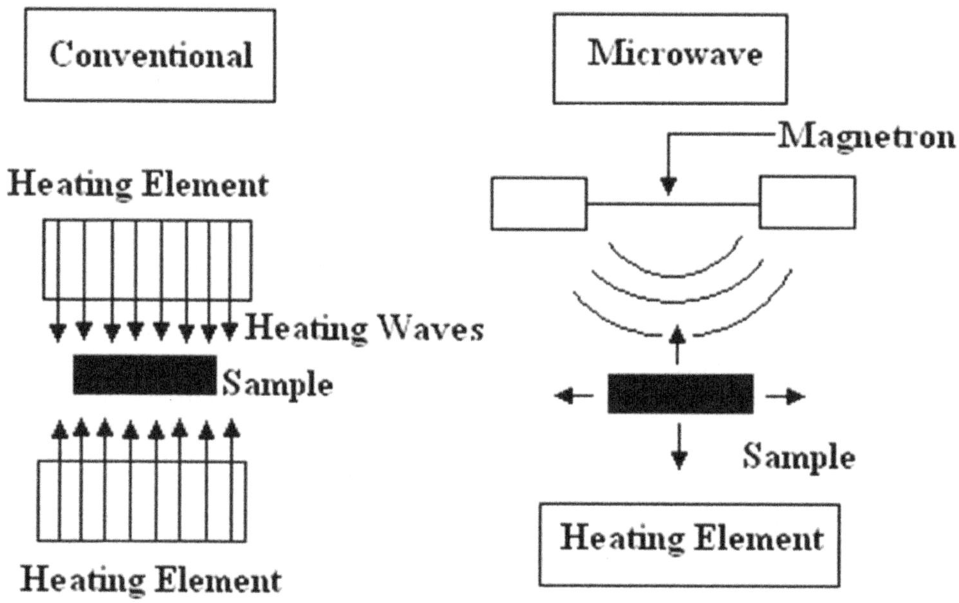

Figure 5.2 Principle of conventional and microwave heating methods.

Therefore, the selective heating process was investigated (Roy & Bhattacharya, 2011, 2012a) and a microwave-assisted synthesis approach was developed which permits the synthesis of nanomaterials at low temperatures, in short time scales, and under cheap and safe reactions conditions (Roy & Bhattacharya, 2012b). Microwave heating leads to the direct interaction between microwaves and materials (Sutton, 1989; Wan, 1993) namely, microwave radiation couples with the material and then is absorbed by the material. The electromagnetic energy is converted to thermal energy. Heat is generated from inside the material, in contrast with conventional heating methods where heat is transferred from outside to inside. This internal heat allows a reduction of reaction time and energy cost and makes new material synthesis possible (Ma *et al.* 1997). Therefore, microwave irradiation as a heating method is significantly faster, simpler and more efficient in energy compared to conventional methods. Consequently, microwave-assisted synthesis method has been widely applied in various fields such as molecular sieve preparation, radiopharmaceuticals, the preparation of inorganic complexes and oxide, organic reactions, plasma chemistry, analytical chemistry, and catalysis (Galema, 1997). Compared with conventional methods, microwave synthesis has the advantages of short reaction time, generating small particle size, and narrow particle size distribution with high purity. Thus, microwave irradiation as a heating method has found numerous applications in synthetic chemistry. Particles of very small sizes show unique physical properties. With the decrease in particle size, an extremely high surface to volume ratio is obtained. This leads to an increase in surface specific active sites for chemical reactions and photon absorption that enhance the reaction and absorption efficiency. The enhanced surface area increases surface states, which changes electron and hole activities and affects the chemical reaction dynamics.

5.2.1.2 Chemical vapor deposition technique

It is a chemical process used to produce high-purity, high-performance solid materials. The process is often used in the semiconductor industry to produce thin films. In a typical CVD process (Figure 5.3), the wafer (substrate) is exposed to one or more volatile precursors, which react and/or decompose on the substrate surface to produce the desired deposit.

Frequently, volatile by-products are also produced, which are removed by gas flow through the reaction chamber.

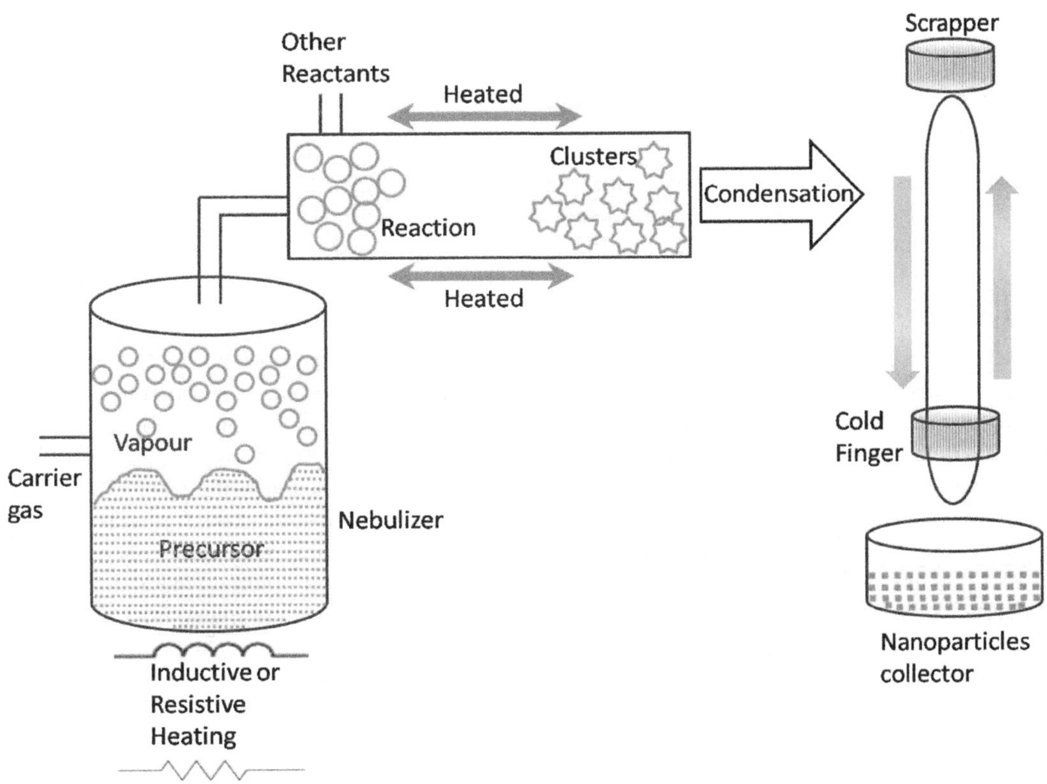

Figure 5.3 A schematic of a chemical vapor deposition reactor.

5.2.1.3 Vapor-phase synthesis technique

In vapor-phase synthesis of nanoparticles, the formation of nanoparticles takes place in gas phase (Figure 5.4). In this synthesis technique the condensation of atoms and molecules is carried out.

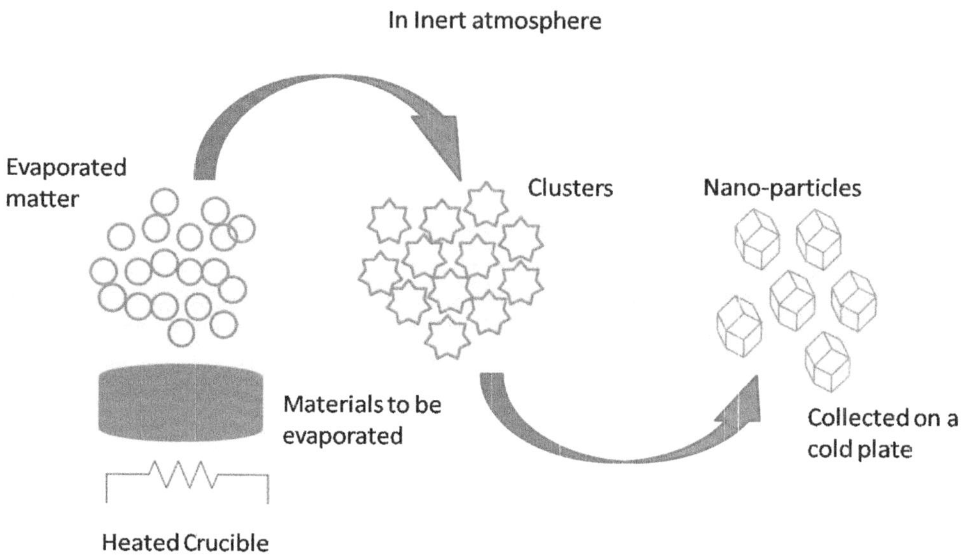

Figure 5.4 Schematic representation of Vapor-phase synthesis of single phase nanomaterials.

The vapor phase synthesis is not new and many multinational companies have been using flame reactors (Figure 5.5) for decades for producing large quantities of nanoparticles. The flame reactors were used for forming various nanoparticles such as carbon black and titanium dioxide.

5.2.1.4 Hydrothermal synthesis technique

Hydrothermal synthesis can be defined as a method of synthesis of single crystals which depends on the solubility of minerals in hot water under high pressure. The crystal growth is performed in an apparatus consisting of a steel pressure vessel called autoclave (Figure 5.6), in which a nutrient is supplied along with water. A gradient of temperature is maintained at the opposite ends of the growth chamber so that the hotter end dissolves the nutrient and the cooler end causes seeds to take additional growth.

Possible advantages of the hydrothermal method over the other types of crystal growth include the ability to create crystalline phases which are not stable at the melting point. Also, materials which have a high vapor pressure near their melting points can also be grown by the hydrothermal method. The method is also particularly suitable for the growth of large good-quality crystals while maintaining good control over their composition. TiO_2 nanowires (Zhang *et al.* 2002), SnO_2 nanoparticles (Chiu & Yeh, 2007), and ZnO nanoparticles (Baruwati *et al.* 2006) were synthesized by hydrothermal method.

5.2.1.5 Micro-emulsion technique

This is relatively a new technique, which allows preparation of ultrafine metal particles within the size range 5 nm <particle diameter <50 nm. The rate of particle nucleation is a function of the percolation degree of micro-emulsion droplets (Figure 5.7). Besides a short introduction into some aspects of the micro-emulsion types and formation, here the kinetics of metallic particle formation using micro-emulsion discussed.

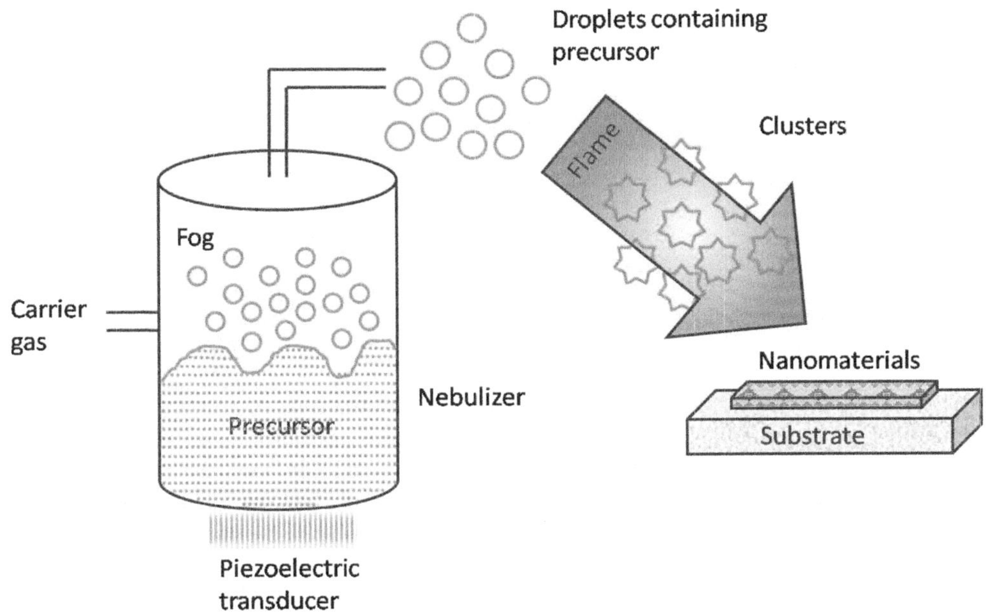

Figure 5.5 Flame assisted synthesis of nanomaterials.

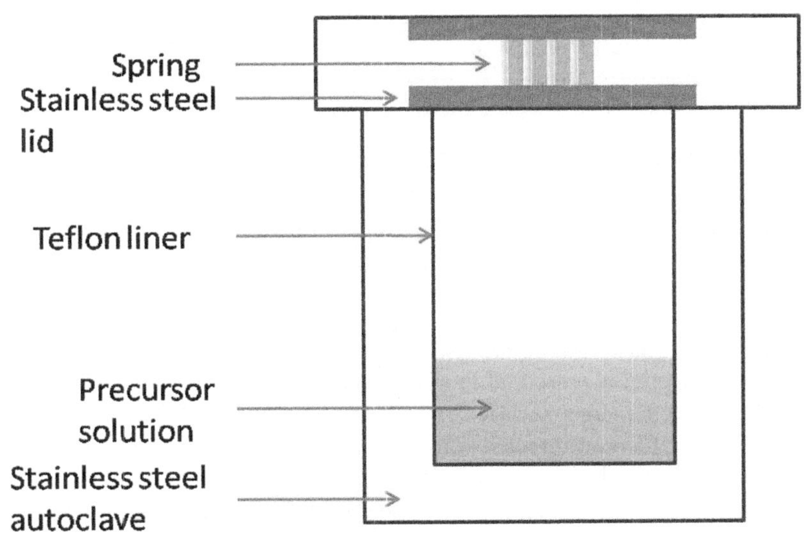

Figure 5.6 Hydrothermal synthesis of nanomaterials (K. Zajaczkowski, Solvothermal Synthesis, 2011).

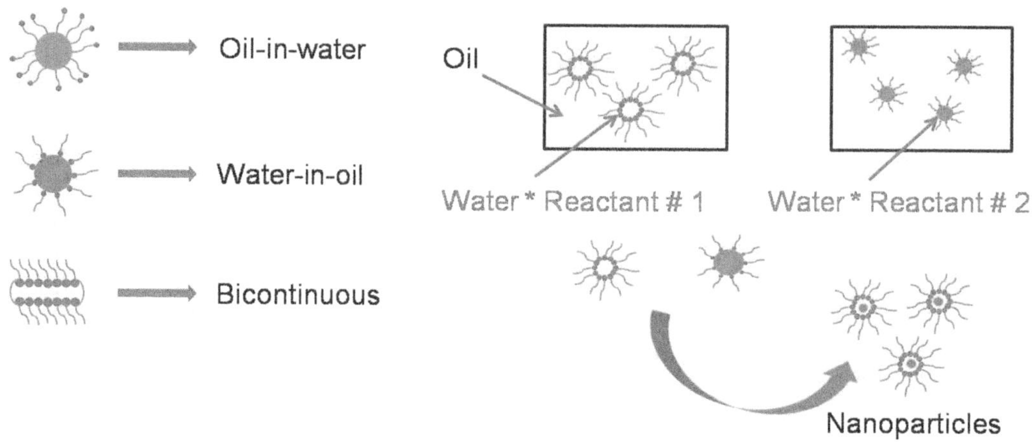

Figure 5.7 Micro-emulsion technique synthesis of nanomaterials.

Effects of stabilizer (emulsifier) type and concentration and the type of continuous phase, reducing agent and additive on the particle formation are summarized and evaluated. The influence of several other parameters such as temperature, the incident light, the nature of metal salts and reaction conditions are also reviewed. These results indicate that the nature of the stabilizer emulsifier, the surface activity of additives and the colloidal stability of micro-emulsion droplets play decisive role on the particle size and distribution during the preparation of metal particles.

5.2.1.6 Sonochemical technique

In this technique, nanoparticles are prepared by irradiating the aqueous or organic dispersion of precursor materials using an ultrasonic probe at room temperature. The size of the particles obtained in this method mainly depends on the solution concentration and the time of sonication. MnO_2 nanoparticles (Zhu *et al.* 2005), CdSe nanomaterials (Zhu *et al.* 2003), and CeO_2 nanparticles (Wang *et al.* 2002) have been synthesized using sonochemical technique.

5.2.2 Physical synthesis method of Nanomaterials

Like chemical synthesis method, physical synthesis method of nanomaterials also contains different techniques which are discussed below.

5.2.2.1 Laser ablation technique

In laser ablation, high-power laser pulses are used to evaporate matter from a target surface such that the stoichiometry of the material is preserved in the interaction. As a result, a supersonic jet of particles (plume) is ejected normal to the target surface. The plume, similar to the rocket exhaust, expands away from the target with a strong forward-directed velocity distribution of the different particles. The ablated species condense on the substrate placed opposite to the target.The ablation process takes place in a vacuum chamber – either in vacuum or in the presence of some background gas. In the case of oxide films, oxygen is the most common background gas. There are many researcher (Amendola & Meneghetti, 2009; Mafuné *et al.* 2002) have been working to synthesized nanomaterials using Laser ablation technique.

5.2.2.2 *Sputtering technique*

A technique used to deposit thin films of a material onto a surface ('substrate'). By first creating gaseous plasma and then accelerating the ions from this plasma into some source material ('target'), the source material is eroded by the arriving ions via energy transfer and is ejected in the form of neutral particles-either individual atoms, clusters of atoms or molecules. As these neutral particles are ejected they will travel in a straight line unless they come into contact with something – other particles or a nearby surface. Saraf *et al.* (1998) reported synthesis of monophase TiO_2 nanoparticles using sputtering technique. Torimoto *et al.* (2006) also elucidated synthesis of highly dispersed ultrafine metal nanoparticles using sputter deposition onto ionic liquids.

5.2.2.3 *Spray route pyrolysis technique*

Nanometer-sized particles (1–100 nm) are of considerable interest for a wide variety of applications, ranging from electronics via ceramics to catalysts, due to their unique or improved properties that are primarily determined by size, composition and structure. A simple, rapid and generalizable aerosol decomposition (spray pyrolysis) process is used for the continuous synthesis of nanoparticles with adjustable sizes, narrow size distribution, high crystallinity and good stoichiometry (Okuyama & Wuled Lenggoro, 2003). Skrabalak and Suslick (2005) reported synthesis of porous MoS_2 and (Panatarani *et al.* 2003) reported synthesis of ZnO nanoparticles using spray pyrolysis.

5.2.2.4 *Inert Gas Condensation technique*

An inert gas condensation technique has been used to prepare nanometer-sized particles of metallic iron by evaporation and agglomeration in a flowing inert gas stream. It is the most advanced technique, although the technique is costly, it is considered as a controlled process as by this method the exact shape and size of the nanoparticle can be synthesized. In inert gas condensation technique, the nanoparticle as soon as they formed rapidly collides with inert gas in a low-pressure environment and thus smaller and controlled nanoparticles are formed (Gracia-Pinilla *et al.* 2010; Pérez-Tijerina *et al.* 2008).

Chapter 6

Experimental techniques

6.1 INTRODUCTION

Synthesis of nanomaterials with desired morphology and composition is one of the most challenging tasks in the field of nanoscience and nanotechnology. In this chapter, an overview is provided on the experimental techniques used for synthesis and characterization of nanomaterials along the lines of the reported research. Understanding the physical characteristics of a material is important in the understanding of their resulting properties. Characterization is particularly important for materials with scale length in the nanometer regime as high degree of changes in material properties take place with the change of dimension. Nanostructured materials that are examined in this work can often have complex structures and present several difficulties in characterization due to the length scales involved. A variety of experimental techniques are used as tools to explore the characteristics of nanomaterials to assist in clarifying their influence on properties. The synthesized different nanostructures need morphological, structural and chemical characterizations for in-depth understanding of the structure-property relationship and their implications in heavy metals remediation technology. The structural and morphological characterization is characterized using X-ray diffraction (XRD), field emission scanning electron microscopy (FESEM) and high resolution transmission electron microscopy (HRTEM). Specific surface area of the synthesized nanoparticles is measured by Brunauer-Emmett-Teller (BET) method. Magnetic properties of maghemite is investigated by Superconducting Quantum Interference Device (SQUID). The chemical precipitation method and Microwave irradiation techniques are used for synthesis of $Ca(OH)_2$, CaO, CaS, and maghemite (γ-Fe_2O_3) nanostructures. Various characterization procedure and the corresponding instruments with the details of specification of various parameters related to the instruments are discussed here. Initial studies focus on the spectroscopic properties of the nanomaterials, while the later studies focus on reactivity employed surface-sensitive techniques that required an ultra-high vacuum environment.

6.2 TECHNIQUE FOR CHARACTERIZATION OF NANOMATERIALS

6.2.1 X-ray diffraction

Radiation striking a material is scattered or absorbed. X-rays, high-energy electrons and neutrons are used to extract structural information of the crystal lattice. Incident radiation of sufficiently smaller wavelengths interact elastically with the regular arrays of atoms in a crystal lattice to yield a diffraction pattern. Both diffraction angles and the intensities in various diffracted beams are functions of the crystalline structure. The diffracted angles depend on the atomic numbers of the constituent atoms and their geometrical relationship with respect to the lattice points. The condition for a crystalline material to yield a discrete diffraction pattern involves the wavelength of incident radiation being comparable to, or less than the interatomic spacing in the lattice. A convenient form of the geometrical relationship determining the angular distribution of the peak intensities in the diffraction pattern from a regular crystal lattice is Bragg's equation 6.2.1

$$2d_{hkl} \sin \theta_{hkl} = n\lambda \tag{6.2.1}$$

where 'n' is an integer referring to the order of reflection, 'λ' is the wavelength of the X-ray used, 'd_{hkl}' is the inter planer spacing of the (hkl) and θ_{hkl} is the angle of diffraction planes. Here it is assumed that the crystals are randomly oriented so that there always exist some oriented crystals so that the Bragg (Jenkins & Snyder, 2012) condition will be satisfied for any set of planes (Figure 6.1).

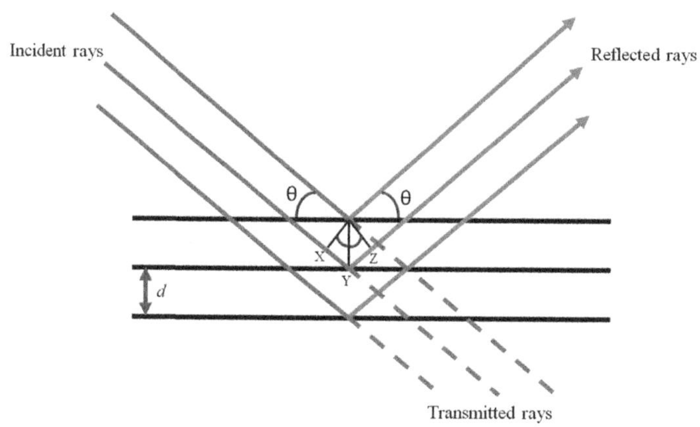

Figure 6.1 Schematic of the X-ray diffraction from lattice.

6.2.1.1 Calculation of crystallite size from X-ray diffraction

Detailed knowledge of crystallite size, shape and strain in a finely divided powder often helps to correlate many physical properties of a system undergoing transformation in a solid-state reaction. X-ray line broadening analysis provides a method of finding bulk average size of coherently diffracting domains and r.m.s. strain. In addition to determining crystal phase, the line widths can be used to estimate the mean crystal size of a sample using the Scherrer (Hammond & Hammond, 2001) equation 6.2.2

$$\beta = \frac{K\lambda}{D \cos \theta} \tag{6.2.2}$$

where β is the line width at half the maximum peak intensity, K is a correction factor for particle shape (0.9 for spheres), D is the crystallite size, and θ is the angle of incidence for the selected diffraction peak. This broadening arises from crystallite sizes that are small enough that insufficient crystal planes are available for the destructive cancellation of reflections that do not occur at exactly the Bragg angle.

6.2.1.2 Density evaluation from X-ray data

The X-ray density of the samples have been computed from the values of lattice parameters using the formula (Smit & Wijn, 1959).

$$d = 4\frac{W}{Na^3} \tag{6.2.3}$$

where 4 represents the number of molecules in a unit cell of a spinel lattice, W is the molecular weight of the sample, N is Avogadro number and a, is the lattice parameter of the sample. The lattice constant for the structure was calculated using the equation:

$$d = \frac{a}{(h^2 + k^2 + l^2)^{1/2}} \tag{6.2.4}$$

6.2.2 Field emission scanning electron microscopy

Scanning electron microscopy (SEM) is a popular and useful tool to study and analyze the morphology of nanoscale materials. In this instrument, thermally excited electrons are used instead of light waves (as in an ordinary microscope) to observe the morphology of the sample under investigation. The electrons are highly energized and hence their de-Broglie wavelength is very small which leads to the high resolution of an electron microscope. Specially constructed electromagnetic lenses are used to focus the electron beam on the sample surface. This facilitates two major benefits of SEM: range of magnification and depth of field in the image, giving 3-D information of image (Figure 6.2). However, in a field emission scanning electron microscope (FESEM), the electrons are generated by application of a very high electric field. An electromagnetic lensing system is used to focus the electron beam. This electron beam, upon impact on the surface of the sample, produces secondary electrons, generated by the scattering from the atoms and electrons of the specimen which are then collected and processed to get the topography of the sample surface of the sample. In the electron gun of FESEM, the cathode provides a narrow electron beam of high as well as low energy. This results in both the improvement in the spatial resolution and minimizes the sample damage due to charging. The incident electron beam on interaction with the solid sample produces secondary ions which are then processed for the imaging. Secondary back scattered electrons are also observed in the solid-electron beam interaction. The electrons, that are emitted from the sample with energy <50 eV are known as secondary electrons and those with energy >50 eV are called back scattered-electrons. In the FESEM, these secondary electrons are used for sample imaging. The electron beam is rastered over the sample surface and the emitted secondary electrons are collected by a detector. The output of the detector is the modulated and processed for imaging purposes.

6.2.3 Energy dispersive X-ray

Energy dispersive X-ray (EDX) is usually used in conjunction with SEM and FESEM. When an incident electron or photon, such as X-ray or γ-ray, hits an atom at a ground state, an electron from an inner electron shell is emitted, leaving a hole or vacancy site in the shell. A more energetic valence electron for the outer shell fills this electron vacancy, resulting in the loss of energy. The excess energy is released in the form of an X-ray emission. The energy of the X-ray emitted is dependent on and unique to the type of elements found in the specimen. The Lithium drifted Silicon (SiLi) detector used in EDX generates a photoelectron when an X-ray strikes it. This photoelectron travels through the Si creating electron-hole pairs. The amplitude of the pulse generated thus depends on the number of electron-hole pairs created, which is itself dependent on the energy of the X-ray impinging on the detector. Hence, the X-ray spectrum can be analyzed for information on the composition of the specimen.

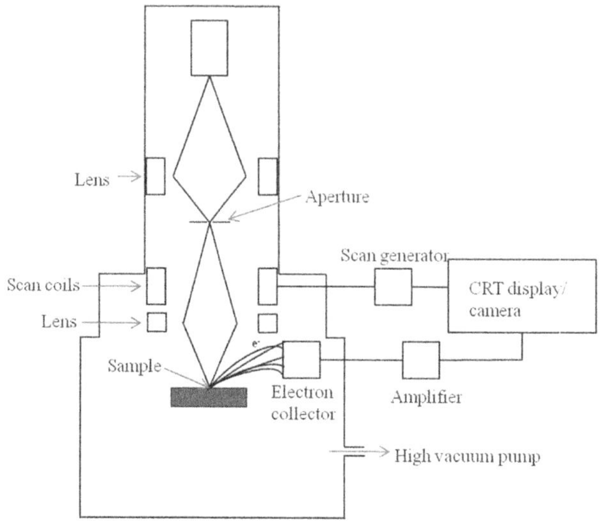

Figure 6.2 Schematic representation of a scanning electron microscope.

6.2.4 Transmission electron microscopy

In transmission electron microscope (TEM), an electron beam is allowed to transmit through a thin layer of specimen. The specimen is supported on a carbon coated copper grid. During transmission through the specimen, the electrons interact with the material. The TEM has three major parts: (i) the illumination system where the electron beam is emitted from the electron gun. The electron beam may be a broad beam or a focused beam. The electron beam is then allowed to pass through the specimen. In Figure 6.3, the part above the sample is part of the illumination system. (ii) The second part is the objective lens and stage. These two items are the most important part of a TEM, called the heart of TEM, and (iii) the third part is the imaging system. It is composed of several intermediate lenses and a projector lens and some connecting devices which produces the image. In the back focal plane of the objective lens, the images of the specimen are formed. The image can be taken in two different modes-the diffraction pattern and the image. When the backfocal plane is taken as the objective plane of the intermediate lens and projector lens, then diffraction patterns will be obtained on the fluorescent screen and the TEM operates in the diffraction mode. This diffraction pattern so produced on the fluorescent screen is equivalent to an X-ray diffraction pattern. If the specimen is single crystalline, the diffraction pattern will appears as linear spots on the fluorescent screen. If it is polycrystalline, then ring-like diffraction patterns appear. Amorphous materials produce a series of diffuse hollos in screen. If the image plane of the objective lens is taken as the objective plane of the intermediate lens and projector lens, an image will form on the screen. The TEM then operates in an imaging mode. The image has some contrast due to several factors: spatial separation between distinct constituent atoms, contrast due to non uniformity of the thickness of the sample, mass contrast and the diffraction contrast which is due to the scattering of the electron waves by various defects in the crystals. Switching from imaging mode to the diffraction mode or the reverse one with just a flick of the switch, makes the practice of TEM possible.

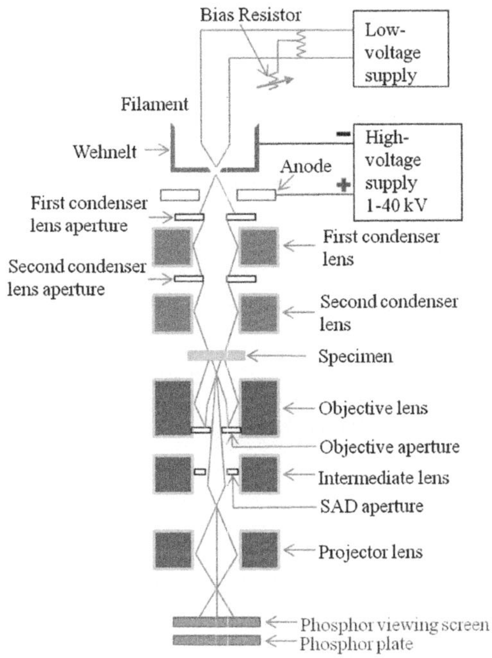

Figure 6.3 Schematic representation of a transmission electron microscope.

6.2.5 Surface area analysis

Surface area is a physical characteristic that is important for materials that are to be used as sorbants. High surface areas imply a larger number of adsorption sites, leading to higher total adsorbate loadings as well as faster sorption kinetics. The most common method used to measure surface area of solid materials is the Brunauer-Emmett-Teller (BET) method (Brunauer *et al.* 1938) and Barrett-Joyner-Halenda (BJH) approach where the sample is cooled to liquid nitrogen temperatures and exposed to a gas adsorbant (typically nitrogen) and the BET equation is applied:

$$\frac{1}{W[(P_0/P)-1]} = \frac{1}{W_m C} + \frac{C-1}{W_m C}\left(\frac{P}{P_0}\right) \tag{6.2.5}$$

where W is the weight of the gas adsorbed at a relative pressure P/P_0 (true equilibrium pressure/equilibrium pressure if no gas were adsorbed), W_m is the weight of a monolayer of adsorbate, and the term C (the BET constant) is related to the energy of adsorption of the first adsorbed layer. When using nitrogen as the adsorbate gas, a linear plot is constructed of $1/W[(P_0/P)-1]$ vs P/P_0 over an appropriate range ($P/P_0 = 0.05 - 0.35$). W_m is obtained from the slope and intercept of the plot, and the specific surface area is calculated using this number and the molecular cross-sectional area of the adsorbate molecule.

6.2.6 Fourier transform infrared spectroscopy

Fourier transform infrared spectroscopy (FTIR) is another widely used infrared (IR) spectroscope for quantitative analyses, where IR radiations pass through a sample. Some of the IR radiation is absorbed by the sample and some of it is transmitted. The resulting IR spectrum represents a fingerprint of the sample with absorption peaks corresponding to the frequency of vibrations between the bonds of the atoms. Because each material is a unique combination of atoms, no two compounds produce the exact same infrared spectrum. Therefore, infrared spectroscopy can result in a positive identification of different kinds of material. In addition, the size of the peaks in the spectrum is a direct indication of the amount of material present. In general the instrument consists of a black body source from where the IR is emitted (Figure 6.4). The beam firstly passes through an aperture, and then enters the interferometer where the spectral encoding takes place. Finally the beam enters the sample compartment where it is transmitted through or reflected off from surface of the sample. This is where specific frequencies, unique to the sample, are absorbed. The beam passes through the detector for final measurements.

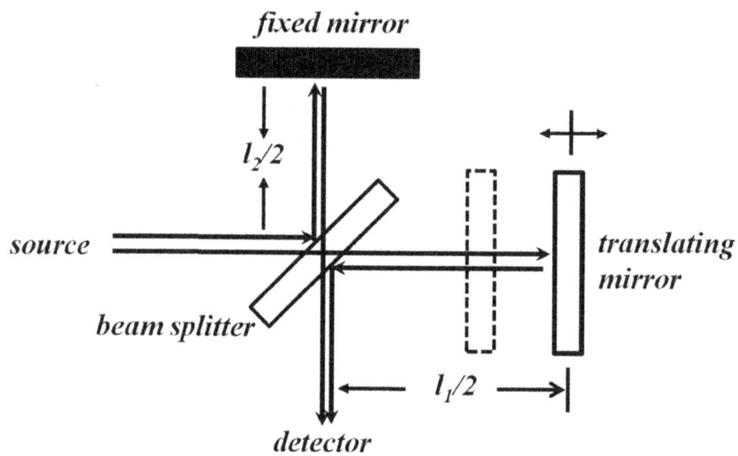

Figure 6.4 Schematic depicting the working of FTIR.

6.2.7 Magnetization measurement

Magnetization (M) is defined as the magnetic moment per unit volume. Experimental measurement of magnetization is required to determine the magnitude of the magnetic moment. It can be measured either isothermally with increasing magnetic field or with changing temperature in a constant magnetic field. Magnetization can be measured by the so-called direct techniques as these consist of direct involvement of magnetic moment in the measurement. These include: (i) force techniques which involve determining the force/torque on a magnetised specimen and (ii) flux techniques which involve either determining the magnetic flux or determining the rate of change of flux linked with the magnetized sample. Force techniques include Faraday balance, torque magnetometer etc., while flux techniques include vibrating sample magnetometer (VSM), Superconducting Quantum Interference Device (SQUID) magnetometer etc. Magnetization can also be measured by various indirect techniques which make use of known relationships between the phenomenon detected and the magnetic properties of the specimen. These include the Hall effect, magneto-optical Kerr effect, Faraday effects, nuclear magnetic resonance, ferromagnetic resonance, Mossbauer effect, neutron scattering, muon spin resonance etc. Each of the methods has advantages and limitations. In the present work, a SQUID magnetometer and a VSM have been used for magnetization measurement. Both of these instruments fall under the category of flux measurement technique.

First, the VSM will be discussed (Figure 6.5). A sample is placed inside a uniform magnetic field, in order to be magnetized, positioned at the end of a long rod holder between the electromagnetic poles. The long holder is connected at its other end to a piezoelectric transducer, producing a periodic vibration. Usually the vibration direction is perpendicular to the magnetic field. The magnetic flux produced around the sample can then be measured by pickup coils, which are sensitive only to the time-dependent magnetic flux. Such a voltage can be sent to a lock-in amplifier and compared with a reference. The measured signal is proportional to the magnetization of the sample. By sweeping the magnetic field, samples become magnetized to different values, and therefore the whole hysteresis loop can be determined. A VSM-SQUID is a new system which benefits from using a SQUID amplifier inside the VSM electronic circuit, increasing the signal to noise ratio and therefore the measurement sensitivity. It also uses a superconducting coil which can produce ± 7 T very quickly. The whole core of this system operates in He vapor on top of LHe, which reduces LHe consumption. It can reach any temperature between 4 and 400 K in a few minutes. This equipment can measure the hysteresis loop or temperature-dependent magnetization for both in-plane and out-of-plane directions. The procedure for zero field cooled (ZFC) and field cooled (FC) measurement is as follows. The sample is cooled initially to a low temperature in the absence of a magnetic field. At low temperature, the magnetic field is applied and measurement is performed up to room temperature. This is the Zero field cooled (ZFC) technique. Again, the sample is cooled in the presence of that magnetic field from room temperature to low temperature and data was taken. This is the Field cooled (FC) technique. The Curie temperatures (TC) are estimated from the plot Magnetization (emu g^{-1}) versus Temperature (K), by extrapolation of linear sections of M (T) up to the intersection with T-axis or from the dM/dT plot.

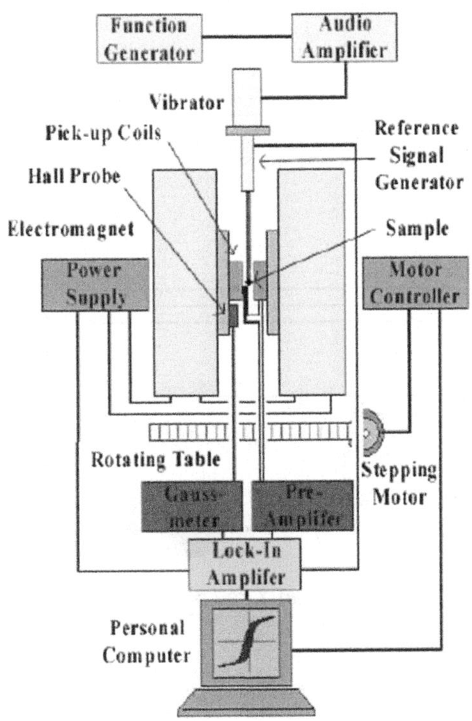

Figure 6.5 Schematic of a VSM.

6.2.8 X-ray photoelectron spectroscopy

X-ray photoelectron spectroscopy (XPS) or electron spectroscopy for chemical analysis (ESCA) is a well-known analytical technique that is used to determine the chemical composition of a surface in addition to the local chemical environment of a given atom. Experiments using XPS require that a sample be under vacuum, where it is bombarded with X-ray photons (200–2000 eV) and relies on the photoelectric effect described by Einstein in 1905 where electrons are ejected from a sample after interacting with the high energy photons. X-ray photons interact with a sample by exchanging enough energy with core-level electrons near the surface of a material to cause electron ejection. Ejected core electrons have a specific kinetic energy (E_k) that is related to the energy of the exciting photon ($h\nu$), the binding energy of the electron to the atom (E_b) and the characteristic work function of the material (Φ) by the relation:

$$E_k = h\nu - E_b - \Phi \qquad (6.2.6)$$

Binding energy is used to determine the oxidation state of elements, and the integrated electron signal can be used to quantify the elemental composition of a surface (Figure 6.6).

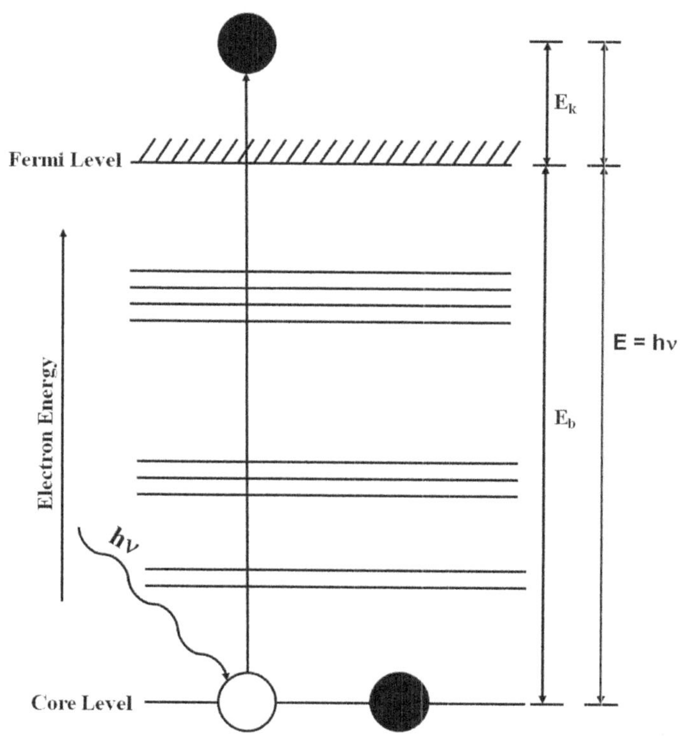

Figure 6.6 Representation of the generation of an X-ray photoelectron.

6.2.9 Zeta potential

Zeta potential, ζ, is based on the mobility of a particle in an electric field and is related to the electrical potential at the junction between the diffuse ion layer surrounding the particle surface and the bulk solution.

A stable suspension requires dominant repulsive forces between particles to maintain dispersion. If the attractive forces dominate, or particles collide with sufficient energy to overcome repulsion, they will begin to flocculate and eventually form agglomerates that settle out from solution. The zeta potential determinations were based on electrophoretic mobility of the nanomaterials in the aqueous medium, which were performed using folded capillary cells in the automatic mode.

6.2.10 Atomic absorption spectrometry

Atomic Absorption Spectrometry (AAS) uses the absorption of radiation by free gaseous atoms in order to achieve quantitative determination of elements (Welz & Sperling, 2008). An atomic absorption spectrometer measures the absorbance which is the logarithm of the rate of incident light power (P_0) to transmitted light power (P):

$$A = \log P_0/P \tag{6.2.7}$$

There is a relationship between P_0 and P when a light beam is decreased by the medium through which it passes:

$$P = P_0 exp(-kL) \tag{6.2.8}$$

where k: absorption coefficient which is a function of wavelength of light, number of atoms in the ground state per unit volume. L: path length in medium

There are two main components in an atomic absorption spectrometer: atom cell which creates atoms at the free gaseous ground state, and optical system to measure the signal. Atom cell dissolvates the liquid sample and dissociates analyte elements into their free gaseous ground state form in which the atoms are available to absorb radiation coming from light source and to create a measurable signal which is proportional to concentration (Haswell, 1991). The atomizer, in which the analyte is atomized, is a flame, graphite tube or quartz tube. In flame atomization, a fixed aliquot of measurement solution is converted into an aerosol in the nebulizer and is transported into the flame which must have enough energy both to vaporize and atomize the sample (Welz & Sperling, 2008).

Chapter 6.1

Case Study: Synthesis of $Ca(OH)_2$ nanoparticles

6.3 INTRODUCTION

Calcium hydroxide ($Ca(OH)_2$) belongs in the category of moderately soluble compounds. Literature on the preparation of moderately water soluble inorganic nanoparticles is rare (Rees *et al.* 1999), compared to the studies concerning nanoparticles of water insoluble compounds (sulfides, oxides, metals, etc.) (Kurihara *et al.* 1983; Lisiecki & Pileni, 1995; Bagwe & Khilar, 1997; Bowers *et al.* 1994; Chhabra *et al.* 1995; Vogel *et al.* 1994). Wilhelmy & Matijević, (1984) reported that the precipitation of metal hydroxide from corresponding salt solution is affected by reaction time and temperature (Matijević & Scheiner, 1978; Sugimoto & Matijevic, 1980). Particularly, above 100°C temperature, formation of nano-scale particles is rapid and sizes are found to be uniform (Sugimoto & Matijevic, 1980). Metal hydroxide precipitation is also affected by concentration of reacting species and type of medium (Hsu *et al.* 1988; Hamada *et al.* 1990). Yura *et al.* (1990) and Hamada & Matijević, (1982) reported that selection of organic solvents can influence size and shape of the particles obtained by the precipitation reaction. The use of diols as solvents in the sol-gel method has also been reported (Wilhelmy & Matijević, 1984; Mingos, 1994). Specifically, ethylene glycol offers two potent advantages, that is, (a) a high dielectric constant, which enhances the solubility of inorganic salts, and (b) a high boiling point (195°C at atmospheric pressure). These properties make it possible to carry out the preparation of inorganic compounds at relatively high temperatures.

6.4 EXPERIMENTAL DETAILS

6.4.1 Synthesis of $Ca(OH)_2$ nanoparticles

$Ca(OH)_2$ nanoparticles were synthesized by a wet chemical method involving dispersion of the metal salt in organic solvent (ethylene glycol) without using surfactant. In a typical run, 25 mL of ethylene glycol (EG) was heated in a conical flask at 150°C, in an oil bath, for 30 min. 12.09 g of $Ca(NO_3)_2{\cdot}4H_2O$ was mixed with the solvent. After dissolving $Ca(NO_3)_2$, 12 mL of NaOH solution (0.7 M) was added dropwise into this mixture while stirring for 10 minutes at 50 rmp. Ca^{2+}–EG complex thus obtained was aged for 5 minutes at room temperature. After that, the supernatant was discarded, and powders were collected by hot vacuum filtration with a 0.45 μm Whatman filter paper. The obtained powders were peptized (Pérez-Maqueda *et al.* 1998; Salvadori & Dei, 2001) with 2-propanol for 2 times, and the resulting suspension was discarded. The peptized powders were then dispersed in 2–propanol in an ultrasonic bath for proper dispersion of the powder. Then the powders were separated by centrifugation at 10,000 rpm for 5 minutes. Still, supernatants contained some agglomeration and were again peptized. The peptization of supernatants was repeated up to five times for low agglomeration. Then the final powder was dried under hot vacuum at 60–70°C for 2 hours.

To study, the effects of different experimental conditions on the final powders, several experiment were performed by changing one parameter at a time, while keeping the others constant. The parameters were (a) aging time of the solution; (b) reaction temperature: 150 and 115°C; (c) concentration of Ca^{2+} in the diol: 0.70, 0.50, and 0.10 mol L^{-1}; (d) concentration of added NaOH: 0.70, 1.50, 0.12 mol L^{-1}. A part of the prepared powder sample (Sample no. 4 with average crystallite size 47 nm) was stocked for 45 days to find out chemical and structural stability of the sample in a capped bottle. A period of 45 days can be considered as normal shelf life.

6.4.2 Characterizations of $Ca(OH)_2$ nanoparticles

The phase analysis and crystal structure of $Ca(OH)_2$ was studied by X-ray diffractograms (XRD). The XRD pattern was recorded using a diffractometer of Philips X-pert Pro X-ray diffractometer (HRXRD), with 0.1789 nm Co–Kα radiation. Particle sizes and morphology were studied with a field emission scanning electron microscope (FESEM) of Carl Zeiss model Supra-40 (with an accelerated voltage 200 kV) and a high-resolution transmission electron microscope (HRTEM) of JEOL (JEM-2100), with an operating voltage 200 kV. The average compositional analysis for $Ca(OH)_2$ were studied with in situ EDS in conjunction with the FESEM imaging of selective regions. In sample preparation for HRTEM studies, a small portion of $Ca(OH)_2$ was dispersed in acetone and sonicated for 30 minutes. Part of this dispersion was dropped over a carbon film supported by a copper grid and dried in vacuum before imaging. The IR spectra of thin pellets of nanoparticles in forms of powders in a KBr matrix were studied. A NexusTM 870 FT–IR (Thermo Nicolet, USA) spectrophotometer equipped with a deuterated triglycine sulfate thermoelectric cool (DTGS-TEC) detector collected the data over a range 500–4000 cm^{-1}.

6.5 RESULTS AND DISCUSSION

Nucleation reactions in the homogeneous phase provide several chemical pathways for obtaining calcium hydroxide nanoparticles. A better control of particle characteristics seems to be offered by judicious choice of counterions of the calcium salt solutions. Counterions of calcium source salts affect the size and the specific surface area of crystallites with a trend that follows a Hofmeister series of anions: (sulfate $<$ chloride $<$ nitrate $<$ perchlorate). The size of particles obtained by the hydrolysis of calcium nitrate in ethylene glycol/aqueous NaOH solution depended on experimental conditions such as reaction temperature, aging time, concentration of NaOH, and concentration of calcium nitrate. Table 6.1 represents conditions under which particles of different size were produced.

Lowering the reaction temperature to 115°C produced larger size particles, and may be due to aggregates of smaller particles. Short aging times resulted in smaller particles, though other conditions remained the same (synthesis No. 4 in Table 4.1). The molar concentration of $Ca(NO_3)_2$ and NaOH also influenced the formation of particles. 25–60 nm particles were obtained at a molar ratio [NaOH]/ [$Ca(NO_3)_2$] equal to 1.4. Increasing this molar ratio resulted larger aggregated particles, but on the lower side, no visible precipitation was observed at a molar ratio of 1.2 (Table 4.1).

Table 6.1 Experimental conditions on average particle size of calcium hydroxide nanoparticles.

Synthesis route	Reaction solvent	Temperature (°C)	NaOH (mol L^{-1})	$Ca(NO_3)_2$ (mol L^{-1})	Aging time (minutes)	Average particle size (nm)
1	Ethylene glycol	150	0.70	0.50	40	80–100
2	Ethylene glycol	115	0.70	0.50	40	>180
3	Ethylene glycol	150	0.70	0.50	25	60–80
4	Ethylene glycol	150	0.70	0.50	5	35–65
5	Ethylene glycol	115	0.70	0.50	5	>100
6	Ethylene glycol	150	1.50	0.50	25	>120
7	Ethylene glycol	150	1.50	0.75	40	60–110
8	Ethylene glycol	150	1.50	0.75	5	–
9	Ethylene glycol	115	1.50	0.75	25	–
10	Ethylene glycol	150	0.12	0.10	40	–

6.5.1 X-ray diffraction of $Ca(OH)_2$ nanoparticles

Figure 6.7 comprises XRD patterns of $Ca(OH)_2$ nanoparticles (a) immediately after synthesis and (b) after 45 days.

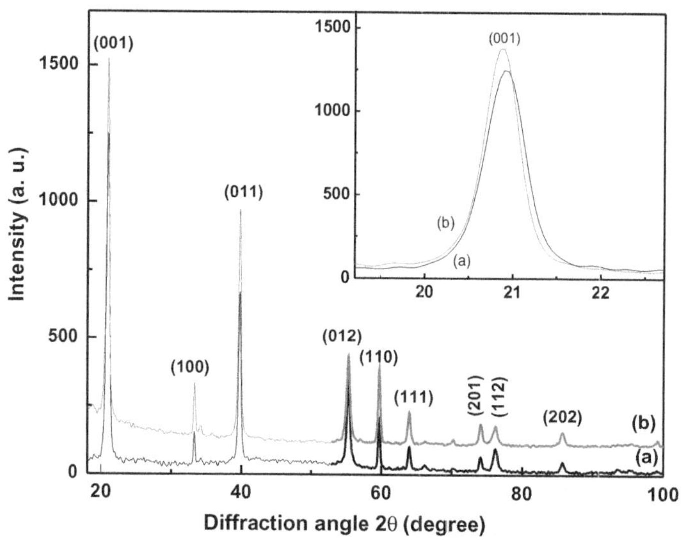

Figure 6.7 The XRD patterns of $Ca(OH)_2$ nanoparticles: (a) immediately after synthesized and (b) after 45 days (In inset a close-up showing a shift of the (001) peak over lower 2θ-value on aging for 45 days). Reprinted from *Synthesis of $Ca(OH)_2$ nanoparticles by wet chemical method*(p-133) by A. Roy, and J. Bhattacharya, 2010, Micro & Nano Letters, Copyright 2010 by The Institution of Engineering and Technology. Reprinted with permission.

As marked by *(hkl)* values, all the peaks in the diffractograms were indexed in terms of a hexagonal crystal structure of $P\bar{3}m1$ space group. Despite little variation in relative peak intensities in the diffractograms, the peak positions (interplanar spacing d_{hkl}) varied due to the variation in the lattice parameters a, and c (Table 6.2), which are similar with the bulk values a = 0.3596 nm and c = 0.4927 nm in $Ca(OH)_2$ from the usual solid state reaction (JCPDS No. 84–1271).

Table 6.2 A comparison in typical structural parameters of synthesized nanoparticles and bulk $Ca(OH)_2$.

Parameter	a (Å)	c (Å)	Average Size (D) (nm)	Volume (V) (nm^3)	Density(ρ) (g cm^{-3})	Surface area (m^2 g^{-1})
Bulk $Ca(OH)_2$	3.5966	4.9278		0.05520	2.22	–
After synthesis	3.5956	4.9277	35	0.05514	2.23	56.84
After 45 days	3.5914	4.9329	46	0.05518	2.22	47.08

The relevant peaks in the XRD pattern for the sample (b) show higher intensities relative to those for (a), suggesting a higher degree of the crystallinity. Due to grain growth from aging for 45 days, the aspect ratio c/a increased by as much as 0.26%, showing an increase in the hexagonality of the lattice, which reveals all feasible lattice reflections in the XRD peaks. The lattice volume V increased (or the density ρ has decreased) regularly from a value 0.05514 nm^3 ($\rho = 2.23$ g cm^{-3}) in (a) to as large value as 0.05518 nm^3 ($\rho = 2.22$ g cm^{-3}), that is, by 0.07%, in the sample (a). The data in Table 6.2 reveals that the cell parameter (c) increased with aging time, indicating c axis growth of crystallite. A close-up of the (001) peak shown in inset of Figure 6.7 clearly demonstrates that it shifted to lower 2θ-values (20.92 for sample (a) to 20.87 for sample (b)) the increasing aging time but shows no splitting of any structural transition. It is also evident from Figure 6.7 that aging time resulted in an increase in average size (D) values. XRD peaks of these samples are broad with large full width half maximum (FWHM), indicating the formation of nanocrystallites. Using XRD peak broadening in the Debye-Scherrer equation 6.2.4 (Yang *et al.* 2006a) determined a 47 nm average crystallite size (D) for sample (a) and that has grown to 57 nm in sample (b).

6.5.2 Microstructure of $Ca(OH)_2$ nanoparticles

FESEM micrographs in Figure 6.8 have nearly hexagonal shapes of synthesized $Ca(OH)_2$ nanoparticles with ~50 nm average particle size on the cross-sections. These micrographs were taken from powders coated with gold on a double sided carbon tape. Compositional analysis, carried out with an EDX analyzer (in conjunction with FESEM) shows no significant impurity present in the sample (Figure 6.8). The observed peaks shown in the inset of Figure 6.8 except Ca and O are obtained with carbon tape. Structural characterization through TEM is a direct way that provides visual demonstration to estimate particle size exactly. Figure 6.9 shows the typical TEM image of the synthesized $Ca(OH)_2$ nanocrystals. Bright field images of the sample indicate that the sample is dispersed nanoparticles having polyhedron shapes with hexagonal projections. The HRTEM image of the nanoparticles is shown in the inset of Figure 6.9, where the lattice fringe is measured to be 0.4927 nm corresponding to the (001) lattice plane of $Ca(OH)_2$ nanocrystals. The images show an abundance of particles whose size distribution is given by the histogram shown in the inset of Figure 6.9, the histogram was obtained by analyzing several frames of similar bright field images using IMAGE J software. The particles have an average size of 35–65 nm, which is in close agreement with an average D-value ~35 nm determined from the XRD peak broadening. Such small particles (separated) present as a large average value of the surface area (Ω) as 52 m^2 g^{-1}. The selected area diffraction (SAD) pattern of the sample is shown in Figure 6.10. The SAD pattern recorded on samples of nanoparticles indicates that they are crystalline in nature. The SAD pattern in Figure 4.4 has six reflections

(001), (100), (110), (111), (011) and (012) at 0.4927, 0.3054, 0.2612, 0.2453, 0.1899, and 0.1788 nm, d_{hkl} values in agreement to the XRD values 0.4927, 0.3104, 0.2632, 0.2463, 0.1932, and 0.1798 nm, respectively.

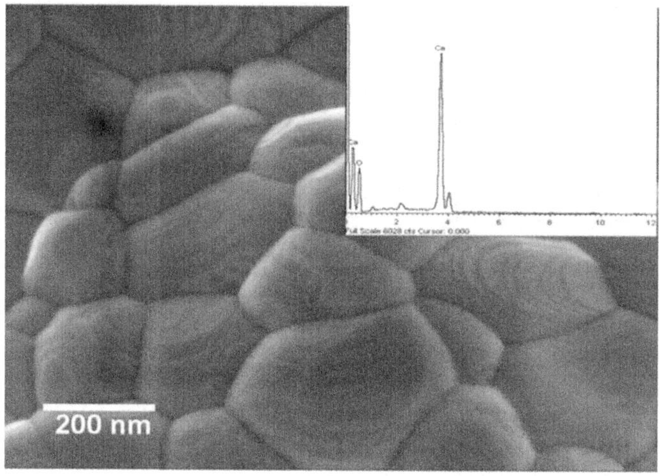

Figure 6.8 FESEM micrograph of synthesized $Ca(OH)_2$ nanoparticles with inset EDX. Reprinted from *Synthesis of $Ca(OH)_2$ nanoparticles by wet chemical method* by A. Roy, and J. Bhattacharya, 2010, Micro & Nano Letters, Copyright 2010 by The Institution of Engineering and Technology. Reprinted with permission.

Figure 6.9 TEM image of synthesized $Ca(OH)_2$ nanoparticles with HRTEM (inset bottom right corner) and particle diameter histogram (inset top right corner). Reprinted from *Synthesis of $Ca(OH)_2$ nanoparticles by wet chemical method* by A. Roy, and J. Bhattacharya, 2010, Micro & Nano Letters, Copyright 2010 by The Institution of Engineering and Technology. Reprinted with permission.

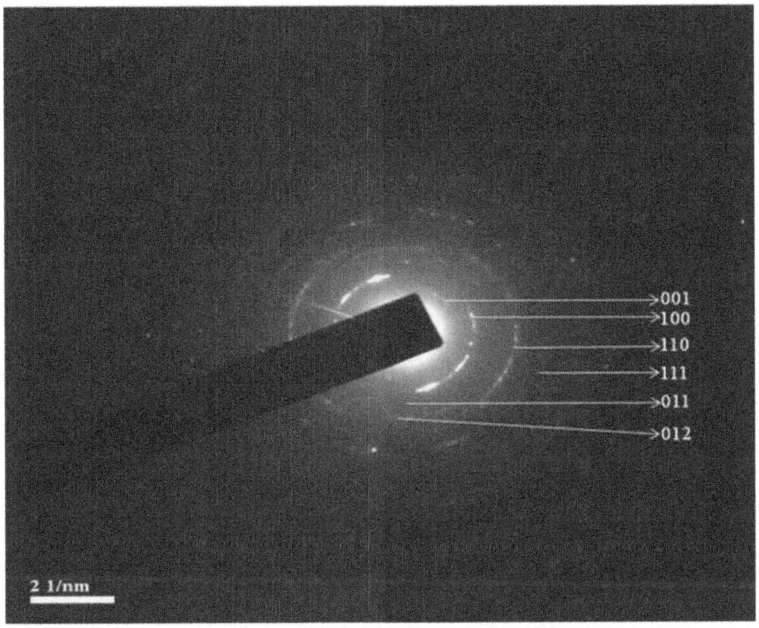

Figure 6.10 Selected area diffraction (SAD) pattern of synthesized $Ca(OH)_2$ nanoparticles. Reprinted from *Synthesis of $Ca(OH)_2$ nanoparticles by wet chemical method*(p-133) by A. Roy, and J. Bhattacharya, 2010, Micro & Nano Letters, Copyright 2010 by The Institution of Engineering and Technology. Reprinted with permission.

6.5.3 FTIR studies of $Ca(OH)_2$ nanoparticles

Figure 6.11 compares FTIR spectra for $Ca(OH)_2$ nanoparticles (a) immediately after synthesis and (b) aged after 45 days. In Figure 6.11, no peaks were observed for metal glycolate in the sample (Tekaia-Elhsissen *et al.* 1989). The peak at 3642.41 cm^{-1} is detected due to the O–H stretching of the solid $Ca(OH)_2$ (Ryskin & Farmer, 1974). A similar FTIR result was reported by Salvadori & Dei, (2001) in $Ca(OH)_2$ nanoparticles prepared by diol as a solvent. In Figure 6.11, peaks of $CaCO_3$ at 876.08 and 1473.34 cm^{-1} were also observed due to the partial carbonation of $Ca(OH)_2$ particle when it was peptized. In comparision with sample (a) and sample (b) intensity for O–H stretching of solid $Ca(OH)_2$ decreased at 3642.14 cm^{-1} wavelength, whereas intensity at 876.08 and 1473.34 cm^{-1} wavelength for $CaCO_3$ increased. The origin of the high reactivity of $Ca(OH)_2$ nanoparticles towards CO_2 resides is greatly dependent on the activity of the large particles surface. Colloidal $Ca(OH)_2$ particles with greater average size synthesized in previous studies (Ambrosi *et al.* 2001; Salvadori & Dei, 2001) showed only partial carbonation. Therefore, the high reactivity of the synthesized particles was observed and attributed to the very small size of particles that not only produces a highly active surface area (52 m^2g^{-1}), but also does not allow the formation of a protective $CaCO_3$ layer on the $Ca(OH)_2$ particles surface. From the XRD pattern of 45 days of $Ca(OH)_2$ nanoparticles, no peak for $CaCO_3$ was observed. It indicates that bulk carbonation did not take place, referring that synthesized $Ca(OH)_2$ nanoparticles are stable up to 45 days in a capped bottle under normal condition.

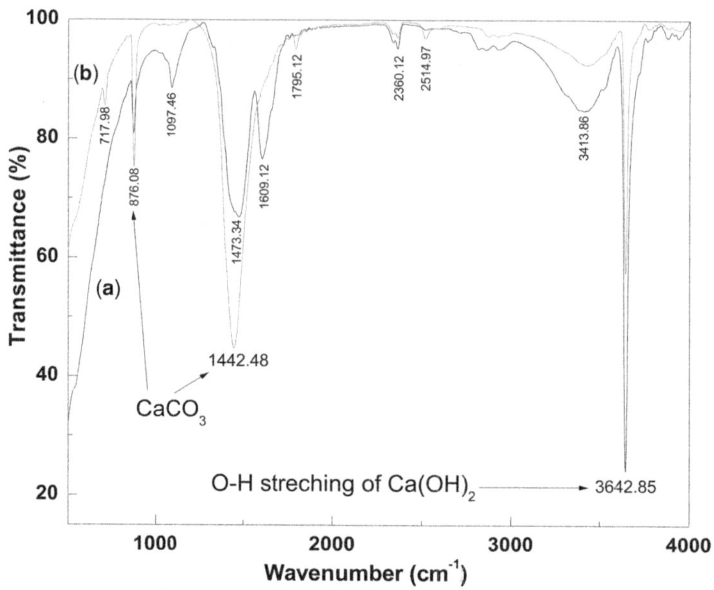

Figure 6.11 FTIR spectra of $Ca(OH)_2$ nanoparticles (a) immediately after synthesized and (b) after 45 days. Reprinted from *Synthesis of $Ca(OH)_2$ nanoparticles by wet chemical method* by A. Roy, and J. Bhattacharya, 2010, Micro & Nano Letters, Copyright 2010 by The Institution of Engineering and Technology. Reprinted with permission.

Chapter 6.2

Case Study: Synthesis of CaO nanoparticles

6.6 INTRODUCTION

Calcium oxide (CaO) is an exceptionally important industrial compound, is used as catalyst, toxic-waste remediation agent, an additive in refractory, in paint as well as for others fundamental applications (Koper *et al.* 1997). Ultra-fine metal oxide particles can be used as bactericide and adsorbent. Particularly CaO has also shown great promise as a destructive adsorbent for toxic chemical agent (Koper *et al.* 1993). Few literatures are found on the preparation of nano-CaO.

There are mainly two methods on the preparation of nano-CaO according to the technical literature. One is thermal decomposition (Bellobono *et al.* 1988, 1991) and other is sol-gel Koper *et al.* (1997). CaO nano-particles can be obtained up to about 14 nm size through the sol-gel method, but the cost is very high. What's more, the process is very complicated and time-consuming. As a consequence, it is very difficult to apply the sol-gel method in industrial applications. The thermal decomposition method has some advantages including simple process, low cost, and ease of obtaining high purity product, etc. So it is a promising and prospective method for application in industry. In the thermal decomposition method, CaO is often obtained directly through calcining $CaCO_3$. A high calcination temperature is needed. In this process, it is very difficult to get nano-scale CaO, but micrometer CaO (above 100 nm), directly through calcination to $CaCO_3$ is possible (Dash *et al.* 2000).

The microwave-assisted route is another method for the synthesis of metal oxides and has been gaining interest in the synthesis of oxide nanomaterials (Palchik *et al.* 2000; Rao *et al.* 1999). Microwave have been used to accelerate organic chemical reactions for some time; because the method is generally rapid, simple, energy efficient and less time consuming (Mingos, 1994; Baghurst & Mingos, 1992). Unfortunately, the exact nature of microwave interactions with reactant during the synthesis of materials is somewhat unclear and speculative. However, energy transfer from microwaves to the material is believed to occur either through resonance or relaxation, which results in rapid heating. Clark and Sutton (1996) have reviewed various aspects of microwave applications reported in the technical literature. Several recent reports have appeared which explain the microwave synthesis method (Wang *et al.* 2003; Liao *et al.* 2003; Jajarmi, 2009; Ai & Jiang, 2009; Jouhannaud *et al.* 2008; Bohnemann *et al.* 2009; Wang *et al.* 2002). Many microwave preparations reported in the technical literature have been prepared on the laboratory scale of only a few grams (Bhat *et al.* 2000; Panneerselvam *et al.* 2001; Kerner *et al.* 2001). However, the use of higher power levels for synthesis of ceramics for specialized applications have also been reported (Nightingale *et al.* 1997; Zhang *et al.* 1997).

6.7 EXPERIMENTAL DETAILS

6.7.1 Synthesis of CaO nanoparticles

$Ca(NO_3)_2{\cdot}4H_2O$ (Marck) and Sodium hydroxide (Marck) and de-ionized water were used and other reagents were obtained from local suppliers. In a typical procedure (Figure 6.12), 0.5 M $Ca(NO_3)_2{\cdot}4H_2O$ and 0.7 M of NaOH were separately dissolved in 50 mL de-ionized water and mixed to form a 100 mL mixture (solution). The mixture was stirred at 50 rpm for 10 minutes at room temperature and it turned into white gels. After 10 minutes stirring at room temperature, the mixture turned into white gels. These were then irradiated by the microwave using a domestic microwave oven having a frequency of 2.45 GHz (maximum power 800 W) at ambient atmosphere for 10 minutes. After microwave processing, the solution was allowed to cool naturally to reach room temperature. The resulting precipitate was collected by vacuum filtration and washed with de-ionized water and absolute ethanol, and dried in a vacuum at 80°C for 1 hour.

$$Ca(NO_3)_2 + 2\,NaOH \xrightarrow[\text{10 min}]{\text{MW Irradiation}} CaO + 2\,NaNO_3 + H_2O$$

2. Vacuum filtration
3. washed with de-ionized water and absolute ethanol
4. dried in a vacuum at 80 °C for 1 h

CaO nanopowder

Figure 6.12 Experimental procedure for synthesis of CaO nanocrystals. Reprinted from 'Microwave-assisted synthesis and characterization of CaO nanoparticles' by A. Roy, and J. Bhattacharya, 2011, International Journal of Nanoscience, Vol 10, No. 3. Copyright @ 2011 by the World Scientific Publishing Company. Reprinted with permission.

6.7.2 Characterizations

The phase analysis and crystal structure of CaO was studied by X-ray diffractograms (XRD). The XRD pattern was recorded using a diffractometer of Philips X-pert Pro X-ray diffractometer (XRD), with 0.1789 nm Co–Kα radiation. Particle sizes and morphology were studied with a field emission scanning electron microscope (FESEM) of Carl Zeiss model Supra-40 (with an accelerated voltage 10–20 kV) and a high-resolution transmission electron microscope (HRTEM) of JEOL (JEM-2100), with an operating voltage 200 kV. The average compositional analysis for CaO was studied with in situ EDS in conjunction with the FESEM imaging of selective regions. In sample preparation for HRTEM studies, a small portion of CaO was dispersed in acetone and sonicated for 30 minutes. Part of this dispersion was dropped over a carbon film supported by a copper grid and dried in vacuum before imaging. The IR spectra of thin pellets of nanoparticles in forms of powders in a KBr matrix were studied. A Nexus^{TM} 870 FT–IR (Thermo Nicolet, USA) spectrophotometer equipped with a deuterated triglycine sulfate thermoelectric cool (DTGS-TEC) detector collected the data over a range 500–4000 cm^{-1}.

6.8 RESULTS AND DISCUSSION

6.8.1 X-ray diffraction studies

The XRD patterns of the synthesized calcium oxide (CaO) are shown in Figure 6.13. The d-spacing values of the sample matched well with the standard PDF database (JCPDS file 77–2376). Unit cell parameters were obtained by least-square refinement of the powder XRD data. The XRD study revealed that the products are monophasic cubic calcium oxide (CaO) with a lattice constant a = 4.801 Å (Fm3m space group) having nanosized particles; the values in the parenthesis indicate the respective Miller indices. The characteristic peaks were higher in intensity and narrower in spectral width, indicating that the products were of good crystallinity.

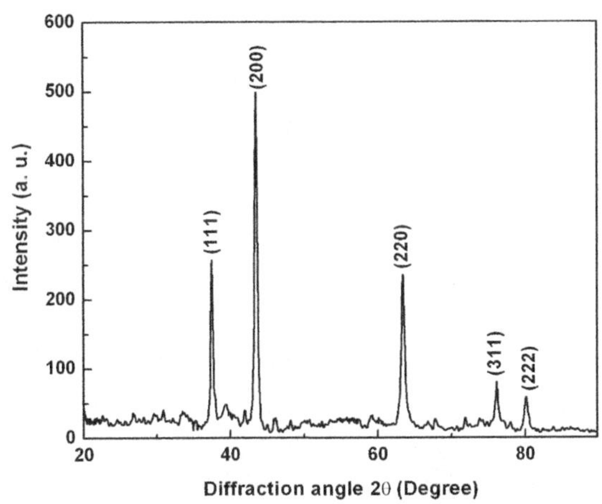

Figure 6.13 The XRD pattern of CaO nanoparticles immediately after synthesis. Reprinted from 'Microwave-assisted synthesis and characterization of CaO nanoparticles' by A. Roy, and J. Bhattacharya, 2011, International Journal of Nanoscience, Vol 10, No. 3. Copyright @ 2011 by the World Scientific Publishing Company. Reprinted with permission.

No peaks corresponding to impurities were detected, showing that the final product is high quality CaO. The crystallite size, density and volume of the samples calculated from XRD data are given in Table 6.3.

Table 6.3 A comparison of typical structural parameters of synthesized nanoparticles and the bulk CaO. Reprinted from 'Microwave-assisted synthesis and characterization of CaO nanoparticles' by A. Roy, and J. Bhattacharya, 2011, International Journal of Nanoscience, Vol 10, No. 3. Copyright @ 2011 by the World Scientific Publishing Company. Reprinted with permission.

Parameter	a (Å)	Average Size (D) (nm)	Volume (V) (nm^3)	Density (ρ) ($g\ cm^{-3}$)	Surface area ($m^2\ g^{-1}$)
Bulk CaO	4.808	–	0.11114	3.34	–
Synthesized CaO	4.801	23	0.11061	3.23	74.46

6.8.2 Microstructure studies

FESEM micrograph in Figure 6.14 have nearly a cubic shape of synthesized CaO nanoparticles with ~27 nm average particle size on the cross-sections. This micrograph was taken from powder coated with gold on a double sided carbon tape. Compositional analysis, performed with an EDX analyzer (in conjunction with FESEM), showed no significant impurity present in the sample (Figure 6.14 inset). The observed peaks in Figure 6.14 except Ca and O could come from carbon tape.

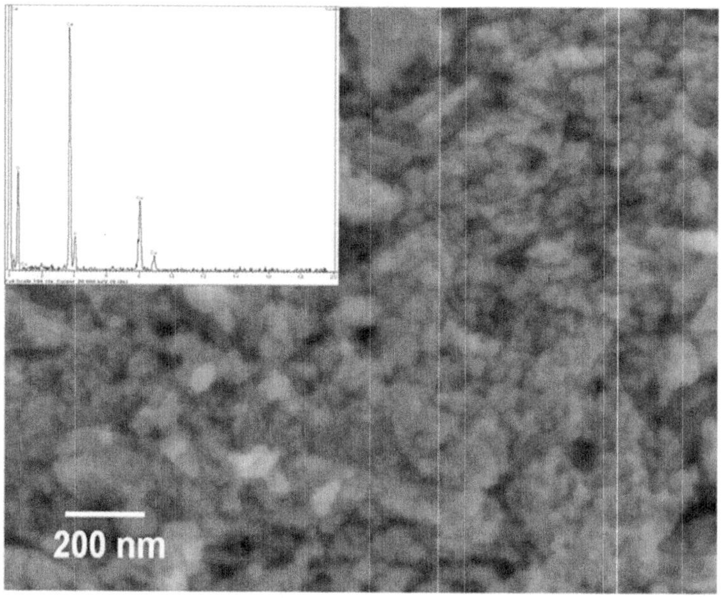

Figure 6.14 FESEM micrograph of synthesized CaO nanoparticles with EDX (inset). Reprinted from 'Microwave-assisted synthesis and characterization of CaO nanoparticles' by A. Roy, and J. Bhattacharya, 2011, International Journal of Nanoscience, Vol 10, No. 3. Copyright @ 2011 by the World Scientific Publishing Company. Reprinted with permission.

Structural characterization through TEM was rather a direct method that provides a visual demonstration to estimate particle size directly. Figure 6.15 shows a typical TEM image of the synthesized CaO nanocrystals. The bright field image of the sample indicates that the sample was dispersive with single-crystal particles having cubic shape. The image shows an abundance of particles whose particles size distribution was given by the histogram shown in the inset of Figure 6.15. The histogram was obtained by analyzing several frames of similar bright field images using IMAGE-J software. The particles have an average size of ~25 nm, which is in close agreement with an average D-value ~23 nm, determined from the XRD peak broadening. Such particles (separated) showed an average surface area (Ω) of 74 $m^2\ g^{-1}$. The HRTEM image of the nanoparticles was shown in the inset of Figure 6.15, where the lattice fringe was measured to be 0.2403 nm corresponding to the (200) lattice plane of cubic CaO. The selected area diffraction (SAD) pattern is shown in Figure 6.16. The SAD pattern is recorded on samples of nanoparticles indicates that they were crystalline in nature. The SAD pattern had five reflections (111), (200), (220), (311), and (222) at d_{hkl} values of 0.2782, 0.2403, 0.1700, 0.1451, and 0.1389 nm in agreement with the XRD values of 0.2771, 0.2400, 0.1697, 0.1447, and 0.1385 nm, respectively.

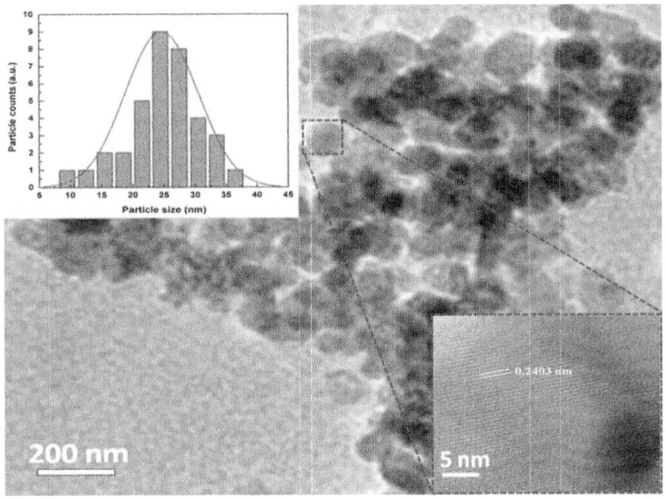

Figure 6.15 TEM micrograph of synthesized CaO nanoparticles with HRTEM (inset bottom right corner) and particle diameter histogram (inset top left corner). Reprinted from 'Microwave-assisted synthesis and characterization of CaO nanoparticles' by A. Roy, and J. Bhattacharya, 2011, International Journal of Nanoscience, Vol 10, No. 3. Copyright @ 2011 by the World Scientific Publishing Company. Reprinted with permission.

Figure 6.16 Selected area diffraction (SAD) pattern of synthesized CaO nanoparticles. Reprinted from 'Microwave-assisted synthesis and characterization of CaO nanoparticles' by A. Roy, and J. Bhattacharya, 2011, International Journal of Nanoscience, Vol 10, No. 3. Copyright @ 2011 by the World Scientific Publishing Company. Reprinted with permission.

6.8.3 Infrared spectroscopy studies

To verify chemically purity, CaO powders were analyzed by FTIR. The bands due to hydroxyl and carbonate were distinctly displayed in the spectrum (Figure 6.17).

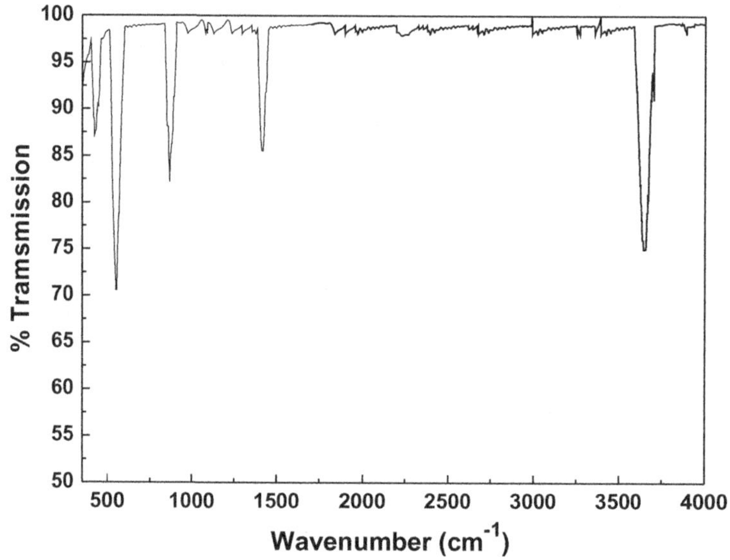

Figure 6.17 FTIR spectra of CaO nanoparticles. Reprinted from 'Microwave-assisted synthesis and characterization of CaO nanoparticles' by A. Roy, and J. Bhattacharya, 2011, International Journal of Nanoscience, Vol 10, No. 3. Copyright @ 2011 by the World Scientific Publishing Company. Reprinted with permission.

The strong band at 3,643 cm^{-1} corresponds to the O–H bonds from the remaining hydroxide (Park *et al.* 2002). Bands at 1,417 cm^{-1} and 866 cm^{-1} correspond to the C–O bond. The wide and strong bands at around 427 cm^{-1} and 553 cm^{-1}, corresponds to the Ca–O bonds (Table 6.4).

Table 6.4 Observed infrared band positions and their assignments. Reprinted from 'Microwave-assisted synthesis and characterization of CaO nanoparticles' by A. Roy, and J. Bhattacharya, 2011, International Journal of Nanoscience, Vol 10, No. 3. Copyright @ 2011 by the World Scientific Publishing Company. Reprinted with permission.

Peak position (cm^{-1})	Assignment	Phase	References
3643	OH	$Ca(OH)_2$	(Nyquist *et al.* 1997; Penel *et al.* 1997)
1476	$\nu_3\ (CO_3)^{2-}$	$CaCO_3$	(Antonakos *et al.* 2007; Penel *et al.* 1997)
870	$\nu_3\ (CO_3)^{2-}$	$CaCO_3$	(Park *et al.* 2002; Penel *et al.* 1997)

Chapter 6.3

Case Study: Synthesis of CaS nanoparticles

6.9 INTRODUCTION

Microwaves produced by magnetrons are electromagnetic waves containing electric and magnetic field components, whose wavelengths lie in the range 1 mm to 1 m (frequency range 0.3–300 GHz). Among them, a large part of the microwave spectrum is used for communication purposes; only narrow frequency windows centered at 0.9 and 2.45 GHz are allowed for microwave heating purposes (Janney, 1988; Link *et al.* 1996; Meek *et al.* 1987; Bykov *et al.* 1997). Microwave heating leads to the direct interaction between microwaves and materials (Sutton, 1989; Wan, 1993) namely, microwave radiation firstly couples with the material and then is absorbed by the material. The electromagnetic energy is converted to thermal energy. Heat is generated from inside the material, in contrast with conventional heating methods where heat is transferred from outside to inside. This internal heat allows a reduction of reaction time and energy cost and makes new material synthesis possible (Ma *et al.* 1997). Therefore, microwave irradiation as a heating method is much faster, simpler and more efficient in energy compared to the conventional methods. Consequently, microwave-assisted synthesis method has been widely applied in various fields such as molecular sieve preparation, radiopharmaceuticals, the preparation of inorganic complexes and oxide, organic reactions, plasma chemistry, analytical chemistry and catalysis (Galema, 1997). Compared with conventional method, microwave synthesis has the advantages of short reaction time, generating small particle size, and narrow particle size distribution with high purity. Thus, microwave irradiation as a heating method has found numerous applications in synthetic chemistry. Particles of very small sizes show unique physical properties. With the decrease in particle size, an extremely high surface to volume ratio is obtained. This leads to an increase in surface specific active sites for chemical reactions and photon absorption that enhance the reaction and absorption efficiency. The enhanced surface area increases surface states, which changes electron and hole activities and affects the chemical reaction dynamics. Alkaline-earth sulfide is a unique phosphor host with a wide band gap and high insulating nature (Tanaka, 1988; Green *et al.* 1988). Possibilities exist for CaS phosphors as emitting materials in electroluminescent display panels, thermoluminescenct dosimeters and cathode ray screens (Xiaolin *et al.* 2001). Because of their outer 3 d electrons, transition metal ions doped luminescence materials are of particular interest for optical applications, especially, Ca^{2+} ions (Anderson *et al.* 2000; Khosravi *et al.* 1995; Konishi *et al.* 2001; Sooklal *et al.* 1996; Yang *et al.* 2002; Yu *et al.* 1996).

6.10 EXPERIMENTAL DETAILS

6.10.1 Synthesis of CaS nanoparticles

$Ca(CH_3COO)_2{\cdot}2H_2O$, ethylalcohol ($C_2H_5OH$) and thioacetamide (TAA) were used for the synthesis of CaS nanoparticles. Milli-Q water was used throughout the experiments. In a typical procedure, an appropriate amount of $Ca(Ac)_2$ was dissolved in 100 mL ethyl alcohol. Then, an appropriate amount of TAA was added into the solution so that the mixture obtains a final concentration of 0.12 M TAA. The concentration of the TAA was slightly higher than $Ca(Ac)_2$ ($S^{2-} > Ca^{2+}$) in order to ensure the reaction gose to completion. Finally, a flask of 250 mL was filled with the mixture solution. The mixture solution was irradiated in a microwave refluxing system for 20 minutes with 20% power (The means of 20% power is that microwave operates in 30 sec. cycle, on for 6 sec., off for 24 sec. The total power is still 650 W). After cooling to room temperature naturally, the precipitate was centrifuged at 4000 rpm for 2 minutes, and then washed with acetone. Heat treatment of the synthesized nanopowder was performed at 100°C for 1 hour under nitrogen atmosphere to obtain well-crystallized calcium sulfide. The final products were collected for characterization.

6.10.2 Characterizations

Microwave oven with 650 W (Sanle general electric corp. Nanjing, China) with refluxing system was used. The XRD pattern was recorded using a diffractometer of Philips X-pert Pro X-ray diffractometer (XRD), with 0.1789 nm Co–$K\alpha$ radiation. Particle sizes and morphology were studied with a field emission scanning electron microscope (FESEM) of Carl Zeiss model Supra-40 (with an accelerated voltage 10–20 kV) and a high-resolution transmission electron microscope (HRTEM) of JEOL (JEM-2100), with an operating voltage 200 kV. The average compositional analysis for CaS were studied with in situ EDS in conjunction with the FESEM imaging of selective regions. In sample preparation for HRTEM studies, a small portion of CaS was dispersed in acetone and sonicated for 30 minutes. Part of this dispersion was dropped over a carbon film supported by a copper grid and dried in vacuum before imaging. The optical absorption spectra were recorded using a UV-vis 2500PC spectrophotometer (Shimadzu Corp., Japan).

6.11 RESULTS AND DISCUSSION

6.11.1 X-ray diffraction studies CaS nanoparticles

Figure 6.18 shows the XRD patterns of CaS nano-powders (a) as-prepared by microwave irradiation for 20 min and (b) after heat treatment. Figure 6.18a shows less defined peaks but in Figure 6.18b strong crystalline reflections appear in the diffraction pattern, and all the six diffraction peaks corresponded to the (111), (200), (220), (311), (222), and (400) planes, indicating synthesis of calcium sulfide cubic crystal structure with space group Fm3m (JCPDS 75–0261, a = 0.5683 nm). The calculated lattice constant of a = 0.5679 nm based on the (200) plane at $2\theta = 36.7°$ is consistent with standard literature value. The average size of the nanoparticles was estimated from the line broadening of the XRD peaks and Scherrer's equation 6.2.2 (Cullity & Stock, 2001). The average particle size was found to be 18 nm. The data in Table 6.5 provides cell parameters including volume, density, and surface area. No impurities are detected by means of XRD analysis.

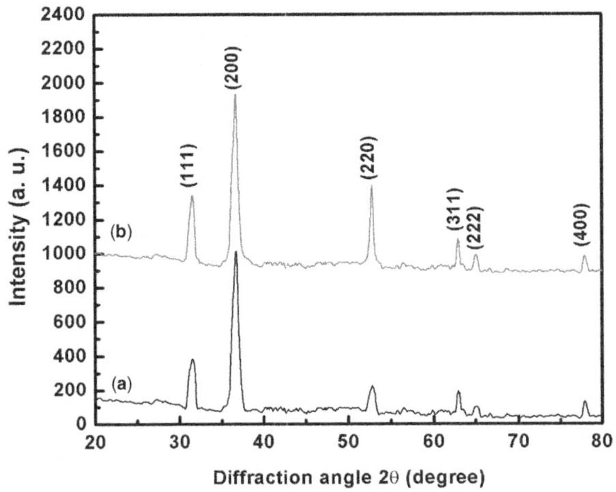

Figure 6.18 The XRD patterns of CaS nanoparticles (a) as-prepared and (b) after heated at 120°C. Reprinted from 'Microwave-assisted synthesis and characterization of CaS nanoparticles' by A. Roy, and J. Bhattacharya, 2012, International Journal of Nanoscience, Vol 11, No. 5. Copyright @ 2012 by the World Scientific Publishing Company. Reprinted with permission.

Table 6.5 A comparison in typical structural parameters of synthesized nanoparticles and bulk CaS. Reprinted from 'Microwave-assisted synthesis and characterization of CaS nanoparticles' by A. Roy, and J. Bhattacharya, 2012, International Journal of Nanoscience, Vol 11, No. 5. Copyright @ 2012 by the World Scientific Publishing Company. Reprinted with permission.

Parameter	a (Å)	Average Size (D) (nm)	Volume (V) (nm^3)	Density (ρ) ($g\ cm^{-3}$)	Surface area ($m^2\ g^{-1}$)
Bulk CaS	5.683	–	0.18354	2.60	–
Synthesized CaS	5.679	18	0.18315	2.61	113.64

6.11.2 Microstructure of synthesized particles

Figure 6.19 shows a typical FESEM image of CaS particles synthesized by this method. It shows that the particles are nearly dispersed and assume a rounded morphology. These micrographs were taken from powder coated with gold on a double sided carbon tape. Compositional analysis, performed with an EDX analyzer (in conjunction with FESEM) showed no significant impurity present in the sample (inset in Figure 6.19). The observed peaks in the inset of Figure 6.19, except Ca and S which came from the carbon tape. The morphology of prepared nanoparticles was studied by TEM. Figure 6.20 shows the typical bright field TEM image of CaS nanoparticles and the shape of these nanoparticles is close to a cubic structure. The HRTEM image of the nanoparticles is shown in inset of Figure 6.20, where the lattice fringe was measured to be 0.28 nm corresponding to the (200) lattice plane of cubic CuS. The image shows an abundance of particles whose particles size distribution is given by the histogram shown in the inset of Figure 6.20; the histogram was obtained by analyzing several frames of similar bright field images using IMAGE-J software. The average size of these nanoparticles is approximately 22 nm, which is in good agreement with the XRD results. Figure 6.21 is the selected-area electron diffraction (SAED) pattern of the nanoparticle taken from its apex. The reflections correspond to (111), (200), (220), (311), (222), and (400) planes of cubic CaS. The SAED pattern and HRTEM image confirmed that single-crystalline CaS nanoparticles were obtained with a uniform distribution.

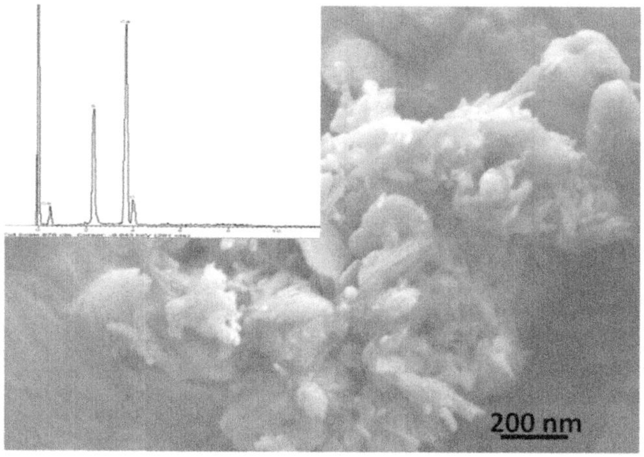

Figure 6.19 FESEM micrograph of synthesized CaS nanoparticles with EDX spectrum (inset). Reprinted from 'Microwave-assisted synthesis and characterization of CaS nanoparticles' by A. Roy, and J. Bhattacharya, 2012, International Journal of Nanoscience, Vol 11, No. 5. Copyright @ 2012 by the World Scientific Publishing Company. Reprinted with permission.

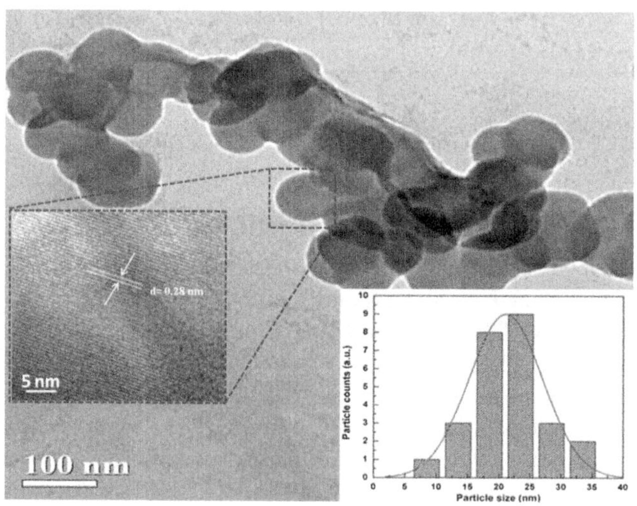

Figure 6.20 TEM image of synthesized CaS nanoparticles with HRTEM (inset middle left) and particle diameter histogram (inset bottom right corner). Reprinted from 'Microwave-assisted synthesis and characterization of CaS nanoparticles' by A. Roy, and J. Bhattacharya, 2012, International Journal of Nanoscience, Vol 11, No. 5. Copyright @ 2012 by the World Scientific Publishing Company. Reprinted with permission.

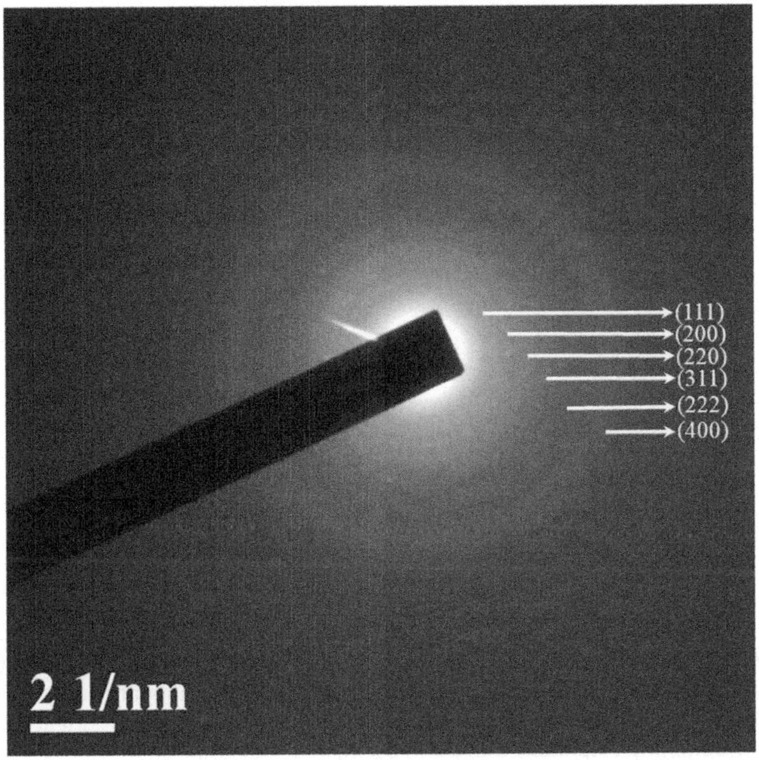

Figure 6.21 Selected area diffraction pattern of synthesized CaS nanoparticles. Reprinted from 'Microwave-assisted synthesis and characterization of CaS nanoparticles' by A. Roy, and J. Bhattacharya, 2012, International Journal of Nanoscience, Vol 11, No. 5. Copyright @ 2012 by the World Scientific Publishing Company. Reprinted with permission.

6.11.3 Optical properties of synthesized particles

Figure 6.22 exhibits UV-vis absorption spectra of the synthesized nanoparticles: (a) as-prepared by microwave irradiation for 20 minutes and (b) after heat treatment. Comparing absorption curve of the samples, it has been found that the absorption spectrum of the heat treatment sample is the same as that the as prepared sample. In the case of heat treatment sample absorption intensity is increased. Each spectrum exhibited a broad shoulder from the onset at about 263 nm. CaS is a wide indirect energy band gap (Eg) semiconductor (Jin *et al.* 2001), and the optical energy gap Eg of the nanophosphors as well bulk phosphors was calculated by the following relation (Pankove, 1971; Sze, 2008):

$$\alpha(h\nu) \sim (h\nu - Eg)^2 \qquad (6.11.1)$$

where $h\nu$ is the photon energy and α is the optical absorption coefficient near the fundamental absorption edge. The absorption coefficients α were calculated from these optical absorption spectra. Figure 6.23 shows the values of $(\alpha h\nu)^{1/2}$ for nanosized CaS plotted as a function of incident photon energy. The indirect energy band gap was obtained by extrapolating the linear portion of the graph and setting $(\alpha h\nu)^{1/2} = 0$. The optical energy band gap for the CaS single crystal (Pankove, 1971) and nanosized CaS are given in

Table 6.6. The increase in the magnitude of the bandgap is indicative of size quantization effects (Zhu *et al.* 2000).

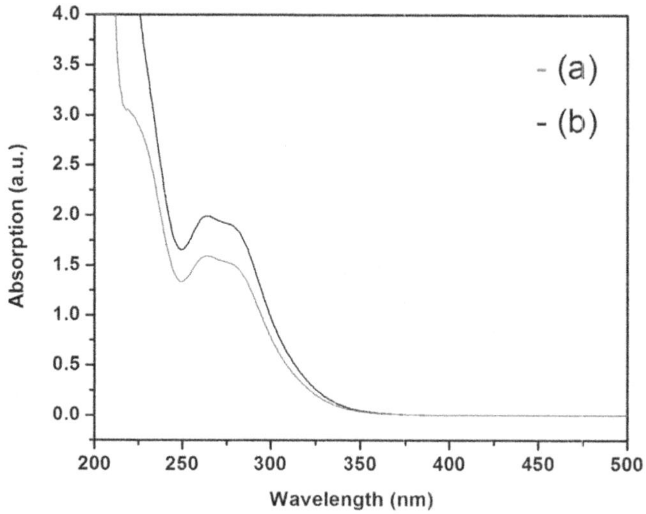

Figure 6.22 UV-Vis spectrum of CaS nanoparticles (a) as-prepared, (b) after heat treatment. Reprinted from 'Microwave-assisted synthesis and characterization of CaS nanoparticles' by A. Roy, and J. Bhattacharya, 2012, International Journal of Nanoscience, Vol 11, No. 5. Copyright @ 2012 by the World Scientific Publishing Company. Reprinted with permission.

Table 6.6 Optical energy band gap for CaS (single crystal) and CaS (nanoparticles). Reprinted from 'Microwave-assisted synthesis and characterization of CaS nanoparticles' by A. Roy, and J. Bhattacharya, 2012, International Journal of Nanoscience, Vol 11, No. 5. Copyright @ 2012 by the World Scientific Publishing Company. Reprinted with permission.

Material	Structure	Indirect energy gap, (eV)
CaS (single crystal)(Pankove, 1971)	Cubic	4.52
CaS (nano)	Cubic	5.00

6.11.4 Mechanism of synthesis of CaS nanoparticles

The phenomenon of producing nanoparticles CaS, due to the high rate of the reactions resulting from the microwave irradiation that provide higher energy, can be explained in the likely reaction pathways. The reactions occurring during microwave irradiation which lead to CaS nanoparticles are understood to be:

$$CH_3CSNH_2 + H_2O \longrightarrow CH_3(NH_2)C(OH)\text{-}SH \qquad (6.11.2)$$

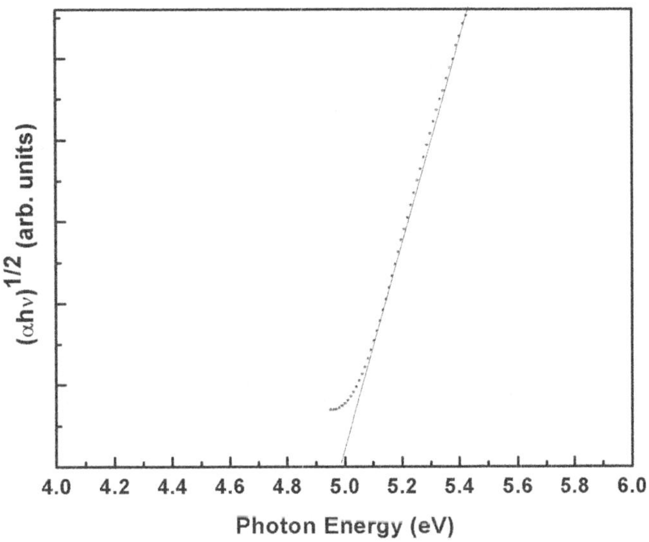

Figure 6.23 Plot for $(\alpha h\nu)^{1/2}$ as a function of the incident photon energy $(h\nu)$ for the CaS nanosized crystallites. Reprinted from 'Microwave-assisted synthesis and characterization of CaS nanoparticles' by A. Roy, and J. Bhattacharya, 2012, International Journal of Nanoscience, Vol 11, No. 5. Copyright @ 2012 by the World Scientific Publishing Company. Reprinted with permission.

$$CH_3(NH_2)C(OH)\text{-}SH + H_2O \longrightarrow CH_3(NH_2)C(OH)_2 + H_2S \quad (6.11.3)$$

$$CH_3(NH_2)C(OH)_2 \longrightarrow CH_3(NH_2)C{=}O + H_2O \quad (6.11.4)$$

$$H_2S + Ca(Ac)_2 \longrightarrow CaS + 2HAc \quad (6.11.5)$$

Equation 6.11.2 indicates that the H_2O reacts with CH_3CSNH_2 to form $CH_3(NH_2)C(OH)$-SH, aided by microwave heating. Repeating this process would then result in formation of $CH_3(NH_2)C(OH)_2$ and H_2S; $CH_3(NH_2)C(OH)_2$ that would immediately lose water to give CH_3CONH_2; Equation 6.11.3 shows the results. Then H_2S further reacts with $Ca(Ac)_2$ to yield nanocrystalline CaS; Equations 6.11.4 and 6.11.5 show the results.

Chapter 6.4

Case Study: Synthesis of γ-Fe_3O_2 nanotubes

6.12 INTRODUCTION

Nanoscale tubular structures have stimulated intensive research interests because of their exceptional physical properties (Iijima, 1991; Appell, 2002).

Iron oxides with nanostructures have attracted a great deal of attention because of their promising properties and applications. Due to their magnetic property, they are widely used in catalysis (Zhang *et al.* 2005), magnetic devices (Zeng *et al.* 2002), environment protection (Wu *et al.* 2005b), sensors (Sun *et al.* 2005), drug delivery (Wu *et al.* 2007) and water splitting (Cesar *et al.* 2006).

Many iron oxide particles with zero-, one-, two- and three-dimensional (0D, 1D, 2D and 3D) nanostructures have been synthesized (Zeng *et al.* 2002; Zhang *et al.* 2005; Wu *et al.* 2005b; Sun *et al.* 2005; Cesar *et al.* 2006; Wu *et al.* 2007). Ferromagnetic nanotubes were considered as candidates for recording heads, biomagnetic sensors, catalysts, etc., because of their expected vortex magnetization state and floatability in liquid as a result of their hollow structure (Goldstein *et al.* 2001; Haberzettl, 2002; Khizroev *et al.* 2002; Sui *et al.* 2004). Iron oxide nanotubes have been synthesized mostly via the so-called template-directed growth method (Shen *et al.* 2004; Liu *et al.* 2005; Wang *et al.* 2006b). However, reports on the template-free synthesis and magnetic properties of γ-Fe_3O_2 (maghemite) nanotubes are relatively scarce (Jia *et al.* 2005; Lv *et al.* 2008; Fan *et al.* 2009; Wu *et al.* 2010). The microwave assisted synthesis rout is gaining significance in the field of nanoparticles synthesis and fabrication. During the last decade, the technique has also been widely applied to prepare inorganic nanostructured materials (Peiró *et al.* 2005; Yin *et al.* 2010; Estruga *et al.* 2010; Bilecka *et al.* 2011; Roy & Bhattacharya, 2010).

6.13 EXPERIMENTAL DETAILS

$FeCl_2{\cdot}4H_2O$, $FeCl_3{\cdot}6H_2O$ and NaOH were all analytical grade. All metal ions solutions were prepared from their nitrate salts (AR) and distilled water. In order to neutralize the anionic charges on the nanoparticles surface, 1 N hydrochloric acid (HCl) of Marck India Co. was used. Distilled water was of Milli-Q quality (resistivity 18 mΩ) (Millipore, USA-Bedford, MD).

6.13.1 Synthesis of γ-Fe_3O_2 nanotubes

In the typical synthesis of γ-Fe_3O_2 (maghemite) nanotubes, 100 mL of 1 M $FeCl_2$ solution was added to 100 mL of 2 M $FeCl_3$. This mixture was stirred under nitrogen atmosphere by nitrogen gas bubbles for 15 minutes. Then, 60 mL of 5 M deoxygenated solution of NaOH was added to the mixture under stirring in ambient atmosphere. To maintain the pH of the final mixture around 11–12, 12.1 N HCl was added. Then the mixture was transferred in a 250 mL round bottom flask with condenser attachment kept in a Multi Synth microwave refluxing system. The maximum deliverable power output was 140 W and was set at a temperature of 180°C for a duration of 20 minutes in the open without the protection of nitrogen or inert gas. The obtained precipitate was separated by a permanent magnet, washed repeatedly with deionized water until neutral, dried in the drying oven at 60°C. The final products were used for all characterization.

6.13.2 Characterization

The phase analysis and crystal structure of as-synthesized samples was studied by X-ray diffractograms (XRD). The XRD pattern was recorded using a diffractometer of Philips X-pert Pro X-ray diffractometer (XRD), with 0.1789 nm Cu–Kα radiation. XPS (X-ray Photoelectron Spectroscopy) was carried out on a PHI 5000 Versa Probe II (Φ ULVAC–PHI, Inc.) model X-ray photoelectron spectrometer instrument with monochromatized Al Ka radiation (1486.6 eV). The pressure inside the analyzer was maintained at 4.2×10^{-7} Pa. The morphologies and microstructures of synthesized samples were characterized with a field emission scanning electron microscope (FESEM) of Carl Zeiss model Supra-40 (with an accelerated voltage 10–20 kV) and a high-resolution transmission electron microscope (HRTEM) of JEOL (JEM–2100), with an operating voltage 200 kV. The composition of synthesized materials was studied with in situ EDS in conjunction with the FESEM imaging of selective regions. The temperature dependence of magnetization was studied and a room temperature magnetic study carried out using a superconducting quantum interference device (SQUID) magnetometer. Surface area, pore size and pore volume were analyzed by Brunauer-EmmettTeller (BET) and Barrett-Joyner-Halenda (BJH) methods using an Autosorb–1(model No. AS1C–9) surface area analyzer (Quantachrome Instrument Corp.). Nitrogen adsorption data were taken at five relative pressures at 77 K to calculate the surface area by BET theory. The surface charge was measured by Zeta–Sizer NanoZS (Malvern) based on electrophoretic mobility of the nanomaterials in aqueous media at different pH. Each measurement was performed in an aqueous solution with a constant ionic strength (in 0.01 M KNO_3), pH was adjusted by adding drops of KOH or HNO_3 solution.

6.14 RESULTS AND DISCUSSION

6.14.1 XRD analysis and XPS analysis

The XRD patterns of the as-prepared products are shown in Figure 6.24a. As shown in Figure 6.24a, the position of the XRD peaks shows good agreement with those of the JCPDS (39–1346) data of the pure cubic phase of maghemite (a = 0.8351 Å). No characteristic peaks of other impurities were observed. More accurately, XPS analysis was performed to determine the composition of the maghemite nanotubes samples Figure 6.24b. The centers of electron binding energy of Fe $2p_{3/2}$ and Fe $2p_{1/2}$ are 709.82 and 723.45 eV, respectively. The shakeup satellite structures at the higher binding energy sides of the main peaks are the fingerprints of the electronic structure of Fe^{2+} ions and indicate that Fe^{3+} ions were absent (Fujii *et al.* 1999). According to the area of the peaks, the ratio of Fe to O in the maghemite nanotubes is about 25:70, which demonstrate that within experimental error a 2:3 Fe/O composition was synthesized.

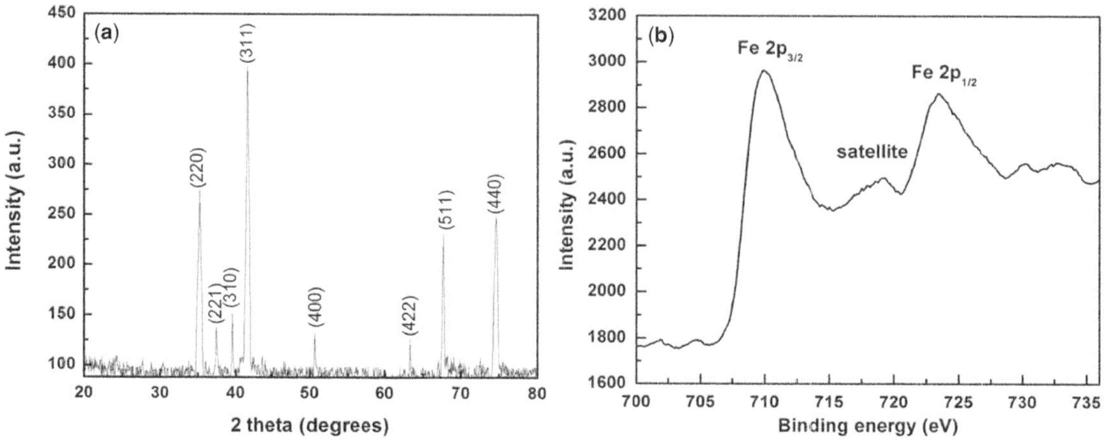

Figure 6.24 The XRD pattern of (a) synthesized nanomaterials and (b) XPS pattern of synthesized nanomaterials. Reprinted from 'Removal of Cu(II), Zn(II) and Pb(II) from water using microwave-assisted synthesized maghemite nanotubes' by A. Roy, and J. Bhattacharya, 2012, Chemical Engineering Journal, Pages No. 493–500. Copyright @ 2012 by Elsevier. Reprinted with permission from Elsevier.

6.14.2 Microstructure studies

The morphologies of the samples were studied by FESEM and HRTEM analysis. Field emission scanning electron microscope (FESEM) imaging shows the morphologies of synthesized material with clear tube-like structure (Figure 6.25a).

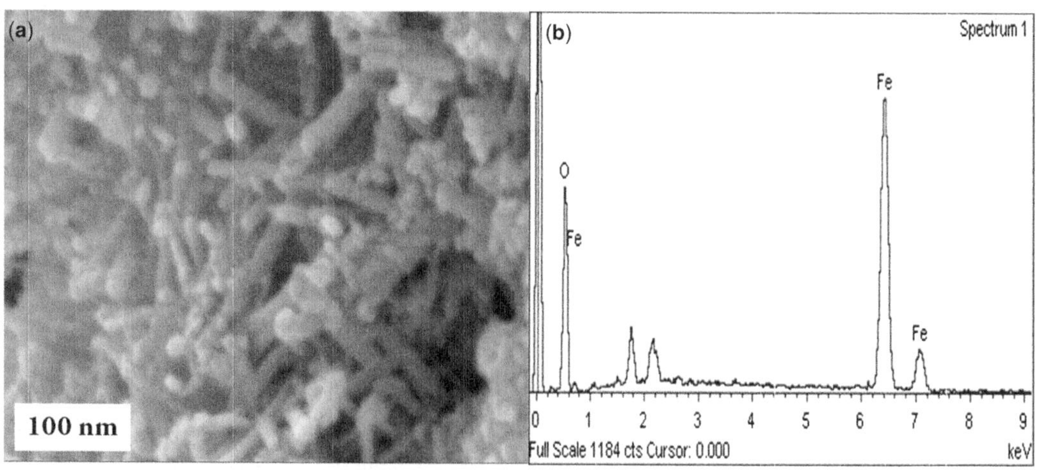

Figure 6.25 FESEM micrograph of synthesized nanomaterials (a) and EDX spectrum of synthesized maghemite nanotubes (b). Reprinted from 'Removal of Cu(II), Zn(II) and Pb(II) from water using microwave-assisted synthesized maghemite nanotubes' by A. Roy, and J. Bhattacharya, 2012, Chemical Engineering Journal, Pages No. 493–500. Copyright @ 2012 by Elsevier. Reprinted with permission from Elsevier.

Compositional analysis, carried out with EDX analyzer (in conjunction with FESEM) (Figure 6.25b) shows no significant impurity present in the sample. The observed peaks in Figure 6.25b except Fe and O probably came from the carbon tape. A high resolution transmission electron microscope (HRTEM) micrograph of maghemite is shown in Figure 6.26. The HRTEM photograph illustrated that the synthesized maghemite were of tube-like shape with outer diameters of ca. 10–15 nm, lengths of 150–250 nm and wall thickness of 6–8 nm. The hollow structure of synthesized nanomaterials is shown in Figure 6.26b.

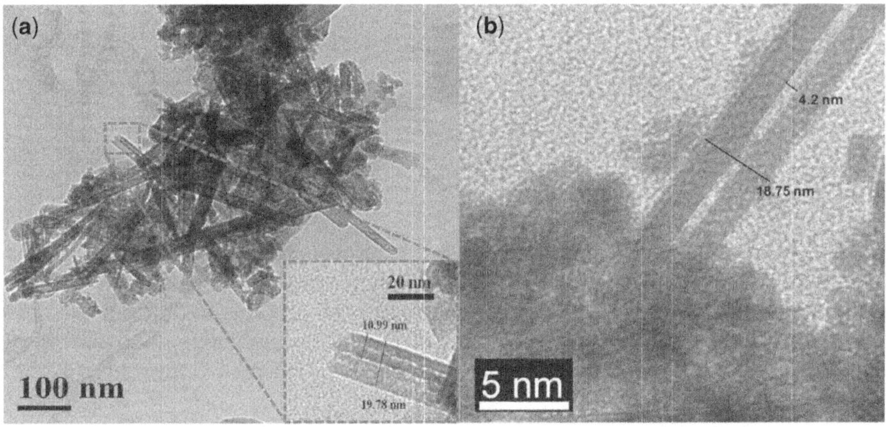

Figure 6.26 TEM micrograph of synthesized nanomaterials (a) and HRTEM image of synthesized maghemite nanotubes (b). Reprinted from 'Removal of Cu(II), Zn(II) and Pb(II) from water using microwave-assisted synthesized maghemite nanotubes' by A. Roy, and J. Bhattacharya, 2012, Chemical Engineering Journal, Pages No. 493–500. Copyright @ 2012 by Elsevier. Reprinted with permission from Elsevier.

6.14.3 Magnetic studies

Magnetic nanoparticles, especially those with special structures, often exhibit unusual magnetic behaviors different from that of bulk solids, owing to finite size effects and microstructure (Bødker *et al.* 2000). To investigate the magnetic properties of synthesized nanotubes, magnetic hysteresis (M-H) loop measurements were carried out in an applied magnetic field at room temperature, with the field sweeping from −50 to 50 kOe. The nanoparticles in aqueous solution showed a tendency of attraction in the presence of a permanent magnet and an aqueous solution of suspended nanoparticles turned transparent near the field within seconds when it was placed nearby (inset of Figure 6.27a). In Figure 6.27a, the saturation magnetizations (M_s) of as–prepared maghemite nanotubes were found to be 68.7 emu g^{-1} at 300 K.

Low coercivity was mainly due to the equiaxed maghemite nanotubes and their anisotropy and the disorder of the crystallographic axis. As, the retentivity (M_r) and coercive force (H_c) were near zero, so one can assume that the synthesized γ-Fe_2O_3 nanotubes were superparamagnetic in nature at room temperature. Figure 6.27(b), shows the temperature dependence of magnetization for the maghemite nanotubes. The curves were acquired between 0 and 300 K using zero-field-cooling (ZFC) and field-cooling (FC) procedures under an applied magnetic field of 100 Oe. It was found that the blocking temperature of the maghemite nanotubes is about 91 K. This result is larger than the blocking temperature of the maghemite nanowires reported, (120 K) (Xiong *et al.* 2003) and (200 K) (Talapin *et al.* 2007), which can be attributed to aligned nanotubes because the blocking temperature increases with the increase of the particle size and degree of alignment (Talapin *et al.* 2007).

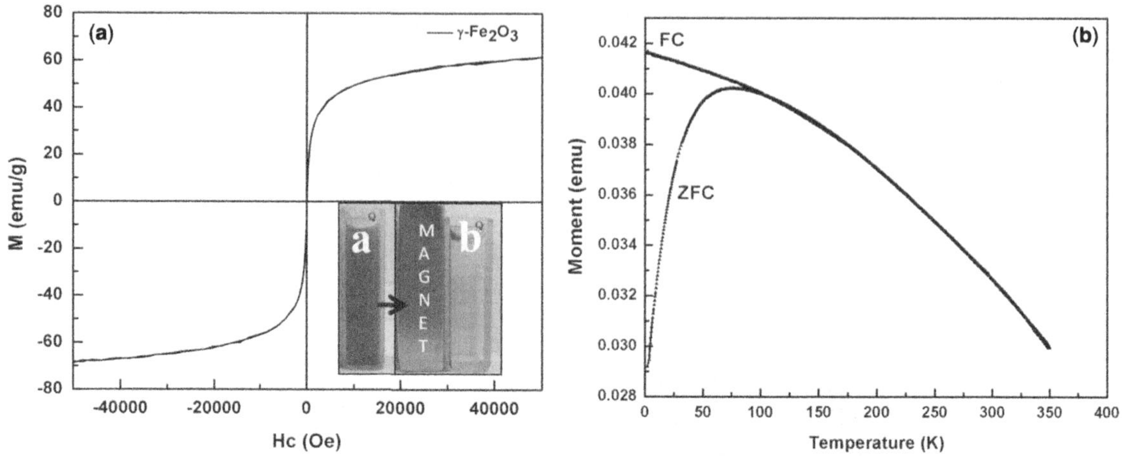

Figure 6.27 M–H loop of maghemite nanotubes (inset diagrams are their corresponding expanded low-field curves and attraction of the nanotubes to a permanent magnet) (a) and ZFC–FC magnetization curves of maghemite nanotubes, measured in a field of 100 Oe (b). Reprinted from 'Removal of Cu(II), Zn(II) and Pb(II) from water using microwave-assisted synthesized maghemite nanotubes' by A. Roy, and J. Bhattacharya, 2012, Chemical Engineering Journal, Pages No. 493–500. Copyright @ 2012 by Elsevier. Reprinted with permission from Elsevier.

6.14.4 Adsorption-desorption isotherms and pore size distribution

The specific surface area of prepared maghemite nanotubes were determined by nitrogen absorption analysis (BET) using a quanta chrome surface area analyzer. Samples were degassed at 150°C for 5 hours under vacuum prior to analysis. The nitrogen adsorption-desorption isotherms of the synthesized nanotube are shown in Figure 6.28. The obtained product showed the type IV isotherms with type H3 hysteresis loops according to BDDT classification (Pierotti & Rouquerol, 1985), indicating the presence of mesopores. The inset in Figure 6.28(b), shows the corresponding pore size distributions of the nanotubes. The nanotubes exhibited a wide pore size distribution ranging from 2 to more than 12 nm. Considering the morphology of the nanotubes observed in Figure 6.28, the smaller pores (~3.57 nm) may correspond to the pores inside the nanotubes and the diameters of these pores are equal to the inner diameter of the nanotubes, while the larger pores (10–12 nm) can be attributed to the aggregation of the nanotubes. It is noteworthy that the one peak in the pore-size distribution at about 3.57 nm is observed in maghemite nanotubes. From the previous HRTEM observation, the inner diameter of the nanotubes was estimated to be in the range of 2–5 nm; therefore, the peak at 3.57 nm reveals the pores due to the inner diameter of the nanotubes. The BET surface area of maghemite nanotubes is quiet large, with a value of 321.638 $m^2\ g^{-1}$ being obtained. Mesopore volumes of maghemite nanotubes was 0.42 $cm^3\ g^{-1}$. Nanotubes with higher surface areas likely lead to higher contact areas to adsorb more heavy metal ions.

6.14.5 Mechanism of synthesis of γ-Fe_2O_3 nanotubes

The reaction between mixtures of $FeCl_2$ and $FeCl_3$ with NaOH under nitrogen or inert atmosphere by conventional heating usually lead to the formation of Fe_3O_4, but in this study, microwave irradiation was used to reach 180°C in place of conventional heating in the presence of atmospheric air. The microwave irradiation increases the reaction rate and as Fe_3O_4 (magnetite) was synthesized, it was readily oxidized to

γ-Fe_2O_3 (maghemite). Similar results by the homogeneous co-precipitation method is reported by Wu & Gao, (2012). Here neither template nor surfactant was employed in the microwave process; it is reasonable to hypothesize that the driving force for the anisotropic growth of maghemite nanotubes is derived from the inherent crystal structure and its chemical potential in solution. Chemical potential influences on the shape evolution of nanocrystals ((Peng & Peng, 2001, 2002). In the case of 1D nanostructure growth, it would be necessary to have a higher chemical potential, controlled by the concentration of aqueous NaOH solution. Long maghemite nantubes with high quality and crystallinity could be obtained in concentrated aqueous NaOH solutions, because higher OH^- ion concentration and higher chemical potential would favor growth of nanotubes.

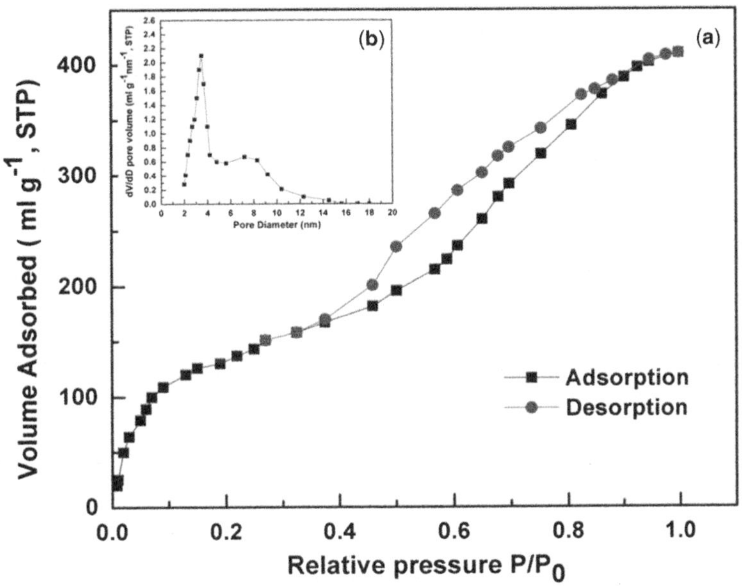

Figure 6.28 N_2 adsorption-desorption isotherms of maghemite nanotubes (a) (inset pore size distribution curve (b)). Reprinted from 'Removal of Cu(II), Zn(II) and Pb(II) from water using microwave-assisted synthesized maghemite nanotubes' by A. Roy, and J. Bhattacharya, 2012, Chemical Engineering Journal, Pages No. 493–500. Copyright @ 2012 by Elsevier. Reprinted with permission from Elsevier.

Chapter 7

Performance of nanomaterials in heavy metals removal

This Chapter discusses the experiments required for batch test to evaluate heavy metals removal efficiency of different nanoparticles synthesized in Chapter 6. A batch test requires all Ca-based nanoparticles to be tested for dissolution in the laboratory. To evaluate magnetic adsorption using γ-Fe_2O_3 nanomaterials batch test is required. Successive batch tests for adsorption and desorption cycle is required to evaluate reusability of γ-Fe_2O_3 nanomaterials.

7.1 PERFORMANCE OF $Ca(OH)_2$, CaO, AND CaS NANOPARTICLES IN HEAVY METALS REMOVAL

7.1.1 Introduction

There are many industrial processes that produce acidic wastewater streams containing low, yet significant concentrations of heavy metals. Electroplating industries have metals concentrations in the range of 100–200 mg L^{-1} in their effluent water (Jeon *et al.* 2001; Cincinnati, 1980). Industries are generally required to adhere to local discharge limits on wastewater concentrations. The heavy metals of particular concern in industrial wastewater are iron (Fe(II)), cadmium (Cd(II)), lead (Pb(II)), zinc (Zn(II)), copper (Cu(II)), cobalt (Co(II)), and nickel (Ni(II)). In chemical precipitation, the dissolved metal ions are converted to the insoluble solid phases (hydroxide or carbonate or sulfide) via a chemical reaction with a precipitant. The resultant precipitate is then separated from the water by sedimentation and/or filtration or flotation (Matis *et al.* 2004; Zamboulis *et al.* 2004). Dissolution characteristics of precipitating agents also influence the heavy metals removal process. The technical literature suggested that due to high surface area to volume ratio, nanoparticles show a higher degree of dissolution in comparison with their respective bulk materials (Meulenkamp, 1998).

Dissolution properties of synthesized Ca-based nanoparticles were investigated via successive batch tests. Factors such as time, pH of the solvent, and dose effects on dissolution of nanomaterials were studied. In batch studies, different factors and their applicability were estimated to obtain optimum results. Building upon knowledge from previous work, a further laboratory investigation was conducted to remove heavy metals from the wastewater using different nanoparticles. Batch tests involved performing tests to investigate the optimization parameters for treatment of heavy metals from synthetic wastewater-focusing primarily on pH and time. To evaluate the removal efficiency of synthesized Ca-based nanomaterials with their bulk materials, dose dependent batch tests were also performed.

7.1.2 Experimental details

7.1.2.1 Materials

Synthesized nanoparticles discussed in Chapter 6, were used in conducting the batch tests. Lime and bulk calcium sulfide were procured from a commercial source. Solutions containing Fe(II), Cu(II), Cd(II), Co(II), Ni(II), Pb(II), and Zn(II) were synthesized from their respective standard reagent grade metal sulfate and nitrate salts (Merck, Germany). Each metal salt, ferrous sulfate heptahydrate ($FeSO_4{\cdot}7H_2O$), copper sulfate pentahydrate ($CuSO_4{\cdot}5H_2O$), cadmium sulfate octahydrate ($CdSO_4{\cdot}8H_2O$), cobalt sulfate heptahydrate ($CoSO_4{\cdot}7H_2O$), nickel sulfate heptahydrate ($NiSO_4{\cdot}7H_2O$), lead nitrate ($Pb(NO_3)_2$), and zinc sulfate heptahydrate ($ZnSO_4{\cdot}7H_2O$), were dissolved in MilliQ water at a concentration of 100 mg L^{-1} of each heavy metal, for precipitation studies and at a concentration of 100 mg L^{-1} of each heavy metal, for the pH dependence study. The pH of heavy metals containing solution was maintained in the range of 1.99–2.02.

7.1.2.2 Solubility of Ca-based nanoparticles and commercial lime and Calcium sulfide

A solubility study of three nanoparticles was conducted in Milli-Q water without addition of any electrolyte. The effect of dose on solubility of nanoparticles was carried out by dissolving known weight (0.1 to 2.0 g L^{-1}) of three Ca-based nanoparticles and commercial lime and calcium sulfide separately, in 250 mL capped flasks containing Milli-Q water and agitated with in magnetic stirrers with 100 rpm at 25°C for 2 hours. The rate of the solubility tests of three Ca-based nanoparticles and lime and calcium sulfide were performed by dissolving a known weight (0.1 g L^{-1}) of three Ca-based nanoparticles and lime in 500 mL capped flasks containing Milli-Q water and agitated with magnetic stirrer at 100 rpm at 25°C for 2 hours. A small aliquot of the sample were collected in predetermined time intervals. The effect of pH on solubility of Ca-based nanoparticles and lime and calcium sulfide was studied by dissolving 0.1 g L^{-1} of each tested material in different pH ranges from 2.0 to 8.0. All capped 250 flasks were agitated in magnetic stirrers at 100 rpm at 25°C for 2 hours. In each experiment, samples were collected and was filtered through 0.45 μm Whatman filter paper and each filtrate was analyzed for the concentration of Ca(II) ion.

7.1.2.3 Dose dependent study for heavy metals removal using $Ca(OH)_2$, CaO nanoparticles and Lime

A dose dependent study was conducted using a known dosage (0.10, 0.25, 0.50, 0.75, 1.00, 1.25, 1.50, 1.75 and 2.00 ($\pm$0.01) g L^{-1}) of $Ca(OH)_2$, CaO nanoparticles, and commercial lime. In 1000 mL beakers, 500 mL of synthetic effluent was treated with each dosage of $Ca(OH)_2$, CaO nanoparticles, and lime separately. Each sample was mixed at 100 rpm by a jar test apparatus for 30 minutes at room temperature. After mixing, samples were allowed to settle for 1 hour. The supernatant was sampled without filtration or other treatment in order to approximate a realistic application. After measuring the pH of each supernatant, the supernatant was digested with concentrated HNO_3 to determine total heavy metal concentrations by atomic absorption spectroscopy (AAS–GBC SensAA). All samples were replicated five times (from the initial stage).

The heavy metals removal capacity of synthesized nanomaterials was calculated according to following equation:

$$\text{Removal efficiency (\%)} = \frac{C_0 - C_t}{C_0} \times 100 \tag{7.1.1}$$

where C_0 and C_t (mg L^{-1}) are initial concentration and the concentration of heavy metal ions at any time t, respectively.

7.1.2.4 Reaction time dependent study for heavy metals removal using $Ca(OH)_2$ and CaO nanoparticles

After determining a suitable dosage of Ca-based nanoparticles in the previous section, similar experimental methodologies were employed for reaction times of 5, 10, 15, 30, 45 and 60 minutes. After being mixed and allowed to settle, the supernatant was sampled and measured for pH, and then digested with concentrated HNO_3 for determining total heavy metal concentrations by AAS. All samples were replicated five times.

7.1.2.5 Sludge volume study

The volume of the decanted sludge was estimated by the volumetric method using Imhoff cones. After 30 minutes of settling, the volume of sludge produced was determined by directly the mL L^{-1} of wastewater treated. The solids were then filtered, dried at 105°C for 24 hours, and weighed. Solid precipitates were characterized by X-ray diffraction analysis.

7.1.2.6 Leaching study

To evaluate the likely performance of precipitates in a landfill environment, the precipitate sludge was tested for its leachabilites following the procedure DIN38414–S4 (Din-Normen, 1984). The vacuum oven dried sludge of 10 g (crushed to <10 mm) was rotated on a roller at 30 rpm with 100 mL of distilled water for 24 hours at an ambient temperature of 25 ± 1°C. The solid/liquid ratio was 1:10. The vessels were centrifuged and the supernatant liquids filtered through a 0.45 μm membrane filter prior to acidification and analysis by AAS. Leaching experiments were performed in duplicate and all analyses were conducted in triplicate, and reported as an average of the values obtained.

7.1.2.7 Dose dependent study for heavy metals removal using CaS nanoparticles and bulk Calcium sulfide

A dose dependent study was conducted using known dosages (0.10, 0.25, 0.50, 0.75, 1.00, 1.25, 1.50, 1.75 and 2.00 (±0.01) g L^{-1}) of CaS nanoparticles and calcium sulfide. In 1000 mL beakers, 500 mL of synthetic effluents were allowed to react with each dosage of CaS nanoparticles and bulk calcium sulfide separately. Each sample was mixed at 100 rpm in a jar test apparatus for 30 minutes at room temperature. After mixing, the samples were allowed to settle for 1 hour. The supernatant was sampled without filtration or other treatment in order to approximate a realistic application. After measuring the pH of each supernatant, the supernatant was digested with concentrated HNO_3 for determining total heavy metal concentrations by AAS. All samples were replicated five times (from the initial stage).

7.1.2.8 pH dependent study for sulfide precipitation using CaS nanoparticles

After determining a suitable dosage of CaS nanoparticles in the previous section, similar experimental methodologies were employed for the pH dependent study with different pH conditions of synthetic wastewater (2.5, 3.5, 4.5, 5.5, 6.0, 6.5, 7.0, 8.5, and 9.5). After being mixed and allowed to settle, the supernatant was then digested with concentrated HNO_3 for determining total heavy metal concentrations by AAS. All samples were replicated five times.

7.1.3 Results and discussion

7.1.3.1 Solubility of Ca-based nanoparticles

$Ca(OH)_2$, CaO, and CaS are known as moderately soluble compounds. The solubility of Ca-based nanoparticles and commercial lime was measured as a function of the dissolved Ca(II) ion. Figure 7.1a shows Ca-based nanoparticles were easily dissolved in water with respect to lime. As particle size was reduced, the dissolution of Ca-based nanoparticles was faster and reached saturated level, within 20 minutes, where as to reach saturated level lime required closer to 60 minutes. From Figure 7.1a, it also observed that there was a trend of a slight decrease in dissolution of Ca-based nanoparticles after reaching highest dissolution. This phenomenon may occur due to aggregation of small particles. The effect of solution pH on solubility of Ca-based nanoparticles and lime is shown in the Figure 7.1b. The solubility of Ca-based nanoparticles and lime were dependent on the pH of the solution. At low pH, high concentration of H^+ elicited dissolution of Ca-based nanoparticles and lime. As the pH increased, the concentration of OH^- also increased and it reduced the dissolution of Ca-based nanoparticles. Figure 7.1c shows the dose dependent dissolution of Ca-based nanoparticles compared to lime. It showed that the concentration of Ca(II) ion increased as the dosage was increased, but the rate of dissolution was higher in the case of Ca-based nanoparticles compared to lime. The optimum dissolution level of Ca-based nanoparticles was reached with a dosage of 1.5 g L^{-1}, whereas to reach the optimum dissolution level of lime, 1.2 g L^{-1} was required. Though $Ca(OH)_2$ nanoparticles have larger particle size than CaO nanoparticles, it was more soluble than CaO nanoparticles. This is because the solubility product (K_{sp}) of CaO is higher than that of $Ca(OH)_2$.

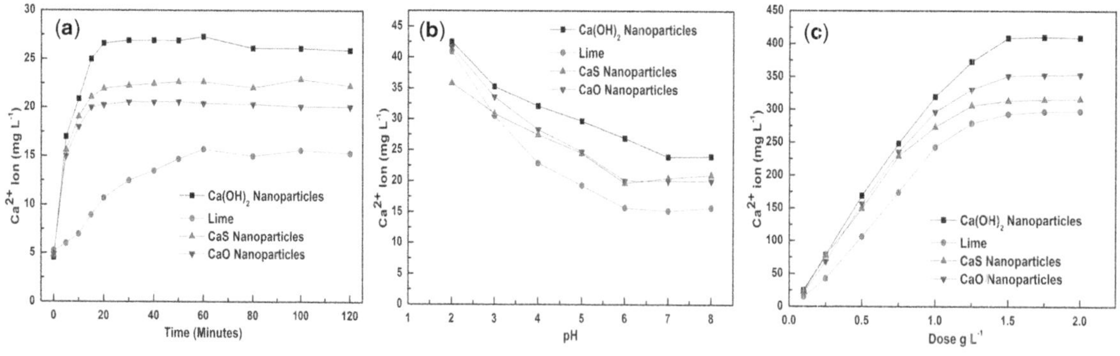

Figure 7.1 Effect of time (a), effect of pH (b), and effect of doses (c) on solubility of Ca-based nanoparticles and lime (Dose 0.25 g L^{-1}, pH 7.0 for 120 minutes).

7.1.3.2 Heavy metal removal

Chemical precipitation is widely used for heavy metal removal from inorganic effluents. After the pH is adjusted to the basic condition, the dissolved metal ions are converted to the insoluble solid phase via a chemical reaction with a precipitating agent. Usually, the precipitates will remain suspended in the water. Therefore, the removal effect may not be satisfactory. In this work, Ca-based nanoparticles and commercial lime were separately added in an incremental way to remove metal ions, until it leveled off at 97.72%–99.84%.

7.1.3.3 Dose dependent study for heavy metals removal using $Ca(OH)_2$ and CaO nanoparticles

The removal of heavy metals from synthetic effluent was investigated using $Ca(OH)_2$ and CaO nanoparticles and lime at nine dosage levels (0.10, 0.25, 0.50, 0.75, 1.00, 1.25, 1.50, 1.75 and 2.00 ($\pm$0.01) g L^{-1}). As nanoparticle dosages were increased, the temperature rose slightly (around 0.2–0.6°C), due to the various complex minerals contained in the synthetic effluents. Tables 7.1 and 7.2 indicate that increasing the dosage of Ca-based nanoparticles caused a rise in pH in the tested effluents. It is clear that the removal of heavy metals gradually increased with increasing pH, until reaching the equilibrium. This effect suggests that Ca-based nanoparticles could be used to raise the solution pH in order to lower the solubility of the metallic constituent, and thus to bring about precipitation. Precipitation with $Ca(OH)_2$ and CaO nanoparticles would constitute various hydroxides. Metal carbonates could also be generated since the reactions of carbonate ions in aqueous solution are the following:

$$Ca(OH)_2 \rightarrow Ca^{2+} + 2OH^- \tag{7.1.2}$$

$$CaO + H_2O \rightarrow Ca(OH)_2 \rightarrow Ca^{2+} + 2OH^- \tag{7.1.3}$$

$$CO_3^{2-} + H_2O \rightarrow HCO^{3-} + OH^- \tag{7.1.4}$$

$$OH^- + M^{2+} \rightarrow M(OH)_2 \downarrow \tag{7.1.5}$$

Because the solubility ranges for the different metals found in the polluted water cover different pH values, and the hydroxide precipitates are amphoteric in nature, the maximum removal efficiency for mixed metals cannot be achieved at a single precipitation pH level (Feng *et al.* 2000). The dose dependent heavy metals removal efficiency of Ca-based nanoparticles are shown in Figures 7.2 and 7.3; those figures showed that the residual metals usually decreased as pH increased.

7.1.3.4 The role of pH on surface charge

The following general equations show the effect of pH on the surface charge of metal oxides and hydroxides (Parks, 1965):

$$MOH_{(surf)} \leftrightarrow MO^-_{(surf)} + H^+(aq) \tag{7.1.6}$$

$$MOH_{(surf)} + H^+(aq) \leftrightarrow MOH^+_{2(surf)} \tag{7.1.7}$$

Thus, the H^+ ion acts as a potential determining ion and effects surface charge. The minimal solubility of the hydroxides of the majority of metals is reached at an approximate pH of 9.5–10.0 (Dyer *et al.* 1998). Table 4.1 presents the values of k_{sp} of various metal hydroxides.

Table 7.1 Heavy metal removal and final pH of solution after treatment with different dosages of $Ca(OH)_2$ nanoparticles.

Dosage ($g\ L^{-1}$)	Final pH	Fe(II) ($mg\ L^{-1}$)		Cu(II) ($mg\ L^{-1}$)		Pb(II) ($mg\ L^{-1}$)		Cd(II) ($mg\ L^{-1}$)		Ni(II) ($mg\ L^{-1}$)		Co(II) ($mg\ L^{-1}$)		Zn(II) ($mg\ L^{-1}$)	
		C_0[a]	C_f[b]	C_0[c]	C_f[d]	C_0[e]	C_f[f]	C_0[g]	C_f[h]	C_0[i]	C_f[j]	C_0[k]	C_f[l]	C_0[m]	C_f[n]
0.10	4.02	99.12	47.17	99.38	38.45	97.80	36.25	97.73	49.53	99.84	45.22	99.82	51.01	99.40	58.40
0.25	4.39	99.14	39.95	99.34	22.47	97.78	29.98	97.70	45.65	99.80	37.55	99.81	47.58	99.40	46.80
0.50	5.09	99.12	33.76	99.41	19.41	97.81	21.36	97.69	35.38	99.89	26.50	99.82	41.21	99.41	32.10
0.75	5.49	99.13	21.91	99.35	16.95	97.82	19.83	97.75	20.96	99.87	20.22	99.80	35.82	99.38	22.30
1.00	6.87	99.16	16.85	99.35	10.23	97.80	10.22	97.74	11.54	99.81	17.56	99.85	23.67	99.39	11.11
1.25	7.89	99.14	10.25	99.32	7.31	97.93	4.01	97.70	5.06	99.85	11.43	99.83	17.78	99.43	6.71
1.50	9.48	99.15	4.46	99.34	1.28	97.91	0.88	97.71	0.47	99.85	6.25	99.87	10.15	99.40	4.19
1.75	10.46	99.11	2.75	99.36	1.03	97.87	1.08	97.76	0.59	99.82	2.21	99.85	5.07	99.45	5.22
2.00	11.26	99.15	1.61	99.4	1.01	97.88	1.87	97.73	0.82	99.83	2.02	99.84	3.02	99.44	6.25

[a]C_0 is initial concentration of heavy metal
[b]C_f is final concentration of heavy metal
[c]C_0 is initial concentration of heavy metal
[d]C_f is final concentration of heavy metal
[e]C_0 is initial concentration of heavy metal
[f]C_f is final concentration of heavy metal
[g]C_0 is initial concentration of heavy metal
[h]C_f is final concentration of heavy metal
[i]C_0 is initial concentration of heavy metal
[j]C_f is final concentration of heavy metal
[k]C_0 is initial concentration of heavy metal
[l]C_f is final concentration of heavy metal
[m]C_0 is initial concentration of heavy metal
[n]C_f is final concentration of heavy metal

Table 7.2 Heavy metal removal and final pH of solution after treatment with different dosages of CaO nanoparticles.

Dosage ($g\ L^{-1}$)	Final pH	Fe(II) ($mg\ L^{-1}$)		Cu(II) ($mg\ L^{-1}$)		Pb(II) ($mg\ L^{-1}$)		Cd(II) ($mg\ L^{-1}$)		Ni(II) ($mg\ L^{-1}$)		Co(II) ($mg\ L^{-1}$)		Zn(II) ($mg\ L^{-1}$)	
		C_0[a]	C_f[b]	C_0[c]	C_f[d]	C_0[e]	C_f[f]	C_0[g]	C_f[h]	C_0[i]	C_f[j]	C_0[k]	C_f[l]	C_0[m]	C_f[n]
0.10	3.12	99.12	47.17	99.38	48.45	97.80	46.25	97.73	53.53	99.84	49.22	99.82	57.01	99.40	58.40
0.25	4.35	99.14	40.95	99.34	27.47	97.78	39.98	97.70	45.85	99.80	39.55	99.81	51.58	99.40	51.80
0.50	4.99	99.12	35.76	99.41	21.41	97.81	25.36	97.69	39.38	99.89	29.50	99.82	46.77	99.41	42.10
0.75	5.77	99.13	30.91	99.35	20.95	97.82	18.83	97.75	19.96	99.87	22.20	99.80	37.82	99.38	30.30
1.00	6.37	99.16	26.85	99.35	18.23	97.80	15.22	97.74	10.54	99.81	19.56	99.85	25.67	99.39	21.10
1.25	7.79	99.14	13.25	99.32	10.31	97.93	10.01	97.70	6.06	99.85	13.43	99.83	20.78	99.43	12.70
1.50	8.68	99.15	7.60	99.34	4.98	97.91	2.98	97.71	3.87	99.85	4.25	99.87	12.15	99.40	8.59
1.75	9.46	99.11	3.15	99.36	1.93	97.87	0.82	97.76	0.59	99.82	3.21	99.85	8.07	99.45	4.22
2.00	10.56	99.15	2.61	99.4	1.21	97.88	1.05	97.73	0.82	99.83	2.02	99.84	4.02	99.44	4.95

[a] C_0 is initial concentration of heavy metal
[b] C_f is final concentration of heavy metal
[c] C_0 is initial concentration of heavy metal
[d] C_f is final concentration of heavy metal
[e] C_0 is initial concentration of heavy metal
[f] C_f is final concentration of heavy metal
[g] C_0 is initial concentration of heavy metal
[h] C_f is final concentration of heavy metal
[i] C_0 is initial concentration of heavy metal
[j] C_f is final concentration of heavy metal
[k] C_0 is initial concentration of heavy metal
[l] C_f is final concentration of heavy metal
[m] C_0 is initial concentration of heavy metal
[n] C_f is final concentration of heavy metal

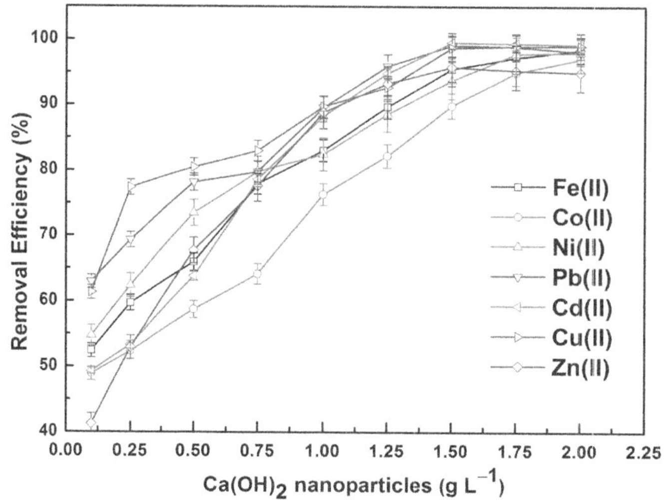

Figure 7.2 $Ca(OH)_2$ nanoparticles dose dependent heavy metal removal.

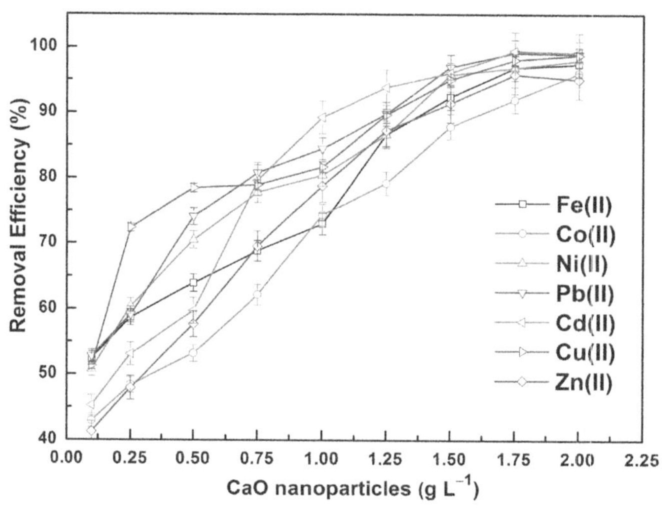

Figure 7.3 CaO nanoparticles dose dependent heavy metal removal.

In the majority of the cases, the formation of the hydroxides, is accompanied by co-precipitation and/or adsorption of metal hydroxides and gives a mixed precipitate (Blais *et al.* 1999; Chu, 1999; Couillard & Mercier, 1992; Marchioretto *et al.* 2005). As pH increases, iron concentration decreases. Fe should be in acid mine drainage mainly as Fe(II) and Fe(II), which should be precipitated at circum pH 8 in accordance to the standard literature (Wei *et al.* 2005). A major reason for the iron precipitation across the studied range may be also due to progressive oxidation of Fe(II) to Fe(II) and its precipitation in the form of $Fe(OH)_3$, that starts from pH 3.5. At pH 11.56 and 10.86 the solubility range of iron is lowered for $Ca(OH)_2$ and CaO nanoparticles, respectively. Cd(II) can be removed by precipitation as hydroxide at pH ranging from 8.0 to 11.0. For the lowest Cd(II) concentrations solubilities at pH above 11.0 was observed for both nanoparticles. Ni(II) and Co(II) were also precipitated as the pH increased above 11.0, lower solubilities was reached. For Cu(II), the lowest solubility range was reached when the pH was around 10.06 using $Ca(OH)_2$, and pH 10.86 with CaO nanoparticles. Although most studies agree that optimum Cu(II) precipitation occurs between pH 9.0–10.3, the most effective treatment was observed at slightly lower operating pH values (Lankford & Eckenfelder, 1990). The lowest solubility range of Pb(II) was achieved with $Ca(OH)_2$ and CaO nanoparticles, when the pH was around 9.68 and 9.56, respectively. Although the optimum pH for Pb(II) precipitation ranged from 6.0 to above 10.0, there was strong evidence that the optimal precipitation was obtained for pH in the range of 9.0–10.0. The amphoteric heavy metals tend to re-dissolve as the pH changes beyond a certain range, presumably also under redox potential changes. Above this pH, lead hydroxide re-solubilized and the concentration of Pb(II) increased as pH increased. At high pH ranges, however, other metals in the wastewater may not be effectively precipitated (Lankford & Eckenfelder, 1990). The lowest solubility range of Zn(II) was reached when the pH was 10.06 and 10.86 for $Ca(OH)_2$, and CaO nanoparticles, respectively. Zn(II) is an amphoteric metal with increasing solubility at both higher and lower pH values (Lankford & Eckenfelder, 1990). As the dose of $Ca(OH)_2$, increased, pH also increased from 10.06 to 11.56 and Zn(II) starts to resolubilize in water.

Cu(II), Pb(II), Cd(II), and Zn(II) reached their lowest solubility, encouraging separation and precipitation at 1.5 g L^{-1} and 1.75 g L^{-1} dose of $Ca(OH)_2$ and CaO nanoparticles respectively, whereas Fe(II), Ni(II) and Co(II) reached their lowest solubility at 2.0 g L^{-1} dose of both nanoparticles. The heavy metals removal performance of nanoparticles was compared with the performance of commercial lime (10–50 μm) under same dosage level. The amount of heavy metals removal by commercial lime is discussed in **Appendix** A.

7.1.3.5 Reaction time dependent study for heavy metals removal

The removal of heavy metals from synthetic effluent using the two selected dosages for two nanoparticles (1.5 and 2.0 g L^{-1} for $Ca(OH)_2$ and 1.75 and 2.0 g L^{-1} for CaO) were examined at six different reaction times (5, 10, 15, 30, 45 and 60 minutes) (Figures 7.4 and 7.5).

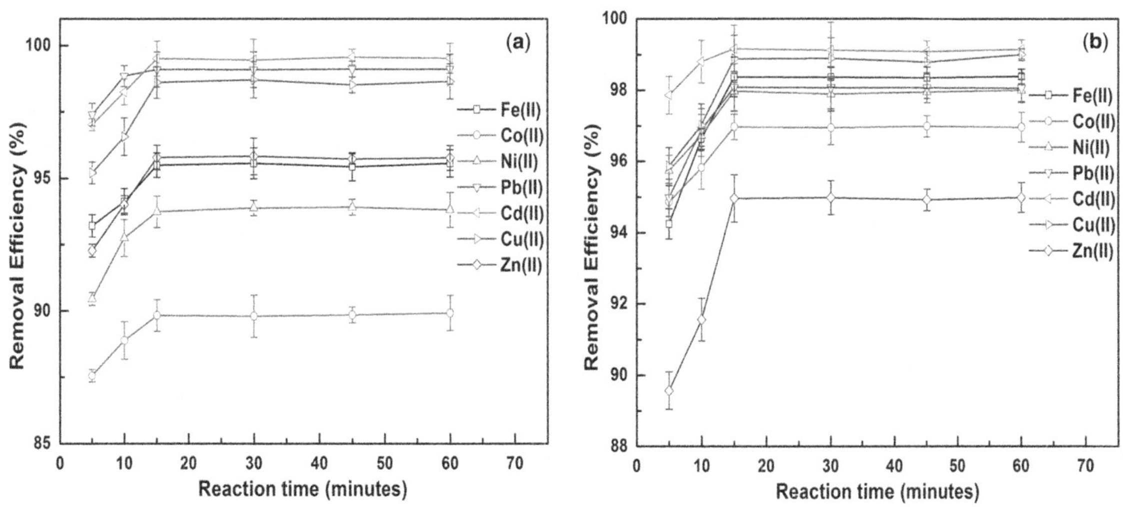

Figure 7.4 The removal efficiency for heavy metals after treatment with 1.5 (a) and 2 (b) g L^{-1} of $Ca(OH)_2$ nanoparticles at different reaction times.

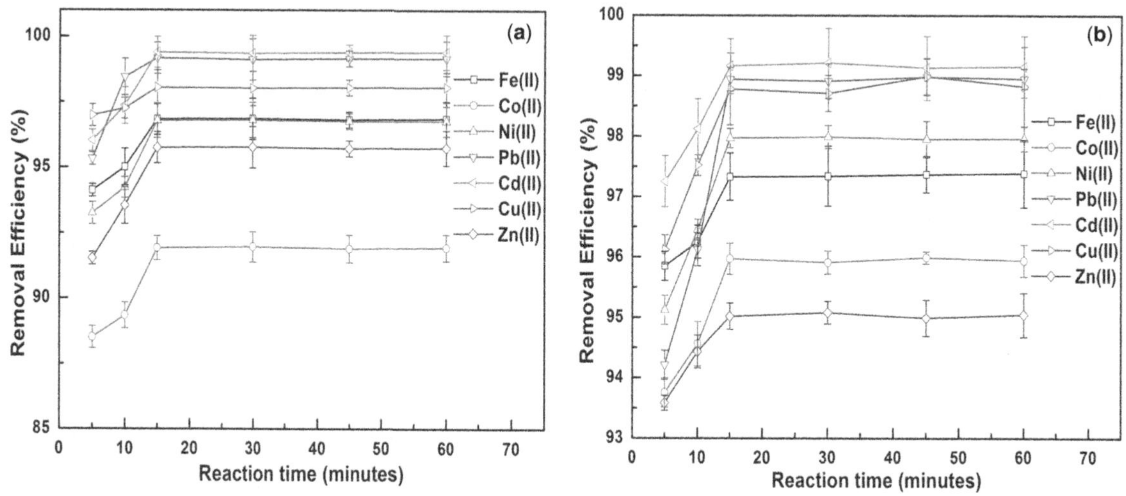

Figure 7.5 The removal efficiency for heavy metals after treatment with 1.75 (a) and 2 (b) g L^{-1} of CaO nanoparticles at different reaction times.

The results reveal that, by increasing the reaction time between the synthetic effluent and the $Ca(OH)_2$ or CaO, an increase in the removal efficiency of most heavy metals occurs until equilibrium is reached. It is clear from the results that the reaction time required for the maximum removal of Fe(II), Cu(II), Pb(II), Cd(II), Ni(II), Co(II), and Zn(II) treated with both selected dosages of CaO was approximately 15 minutes. For $Ca(OH)_2$, the optimal reaction times for maximum removal were 15 minutes for all heavy metals, for both 1.5 and 2.0 g L^{-1} dosages. The results for Pb(II), however, differed for the dosages. According to these results, a large number of heavy metals were removed when the equilibrium reaction time was at 15 minutes. This time length permitted the settling of these wastes, producing a clear supernatant. A 15 minutes reaction time was therefore selected for further study.

From the Figure 7.1b one can observe that to reach complete dissolution of $Ca(OH)_2$ and CaO nanopaticles, 15–20 minutes were required. In spite of other cations being present in the effluents both nanoparticles were available for reactions and reached equilibrium. The efficiency of removing the majority heavy metals decreased when the reaction time exceeded 15 minutes indicating high reaction rates, leading to decreased concentrations. This effect could be attributed to the reaction time being prolonged beyond a point without any noticeable increase in removal. Beyond this point, efficiency could go down and levels off, as contact times get longer. This may also be the result of a decreasing strength of sorption and/or the breaking of the sorbate–surface bond (El-Awady & Sami, 1997). A research study by Namasivayam and Ranganathan (1995) on the removal of Cd(III) from wastewater by adsorption on waste Fe(III)/Cr(III) hydroxide found that by agitation the cadmium uptake could be increased. In our experiments, the uptake remained constant, after a time to achive equilibrium of 30 minutes.

7.1.3.6 Selected treatment conditions study for heavy metals concentration

From the previous studies described in Section 7.1.2.3, the selected dosages of $Ca(OH)_2$ nanoparticles were 1.5 and 2.0 g L^{-1} and CaO nanoparticles were 1.75 and 2.0 g L^{-1}, and the selected reaction time (from

Section 7.1.2.4) was 15 minutes. Four conditions were thus chosen for use in the next stages of the research (Table 7.3).

Table 7.3 The results of heavy metals concentration after treatment at selected conditions.

Parameter	Concentration ($mg\ L^{-1}$)	Concentration[a] ($mg\ L^{-1}$) after treatment				EPA limit ($mg\ L^{-1}$)
		Condition 1[b]	Condition 2[c]	Condition 3[d]	Condition 4[e]	
pH Heavy metals	1.99–2.12	9.48–9.58	11.26–11.08	9.46–9.54	10.56–10.45	6.0–9.5
Fe(II)	99.13	4.45	1.62	3.13	2.62	3.0
Cu(II)	99.36	1.37	1.10	1.95	1.16	3.0
Pb(II)	97.84	0.87	1.89	0.83	1.03	0.1
Cd(II)	97.72	0.47	0.85	0.60	0.81	2.0
Ni(II)	99.84	6.15	2.04	3.22	2.02	3.0
Co(II)	99.83	10.13	3.02	8.07	4.04	–
Zn(II)	99.41	4.20	5.01	4.23	4.93	5.0
Sludge volume ($mL\ L^{-1}$)		72	100	68	88	–

[a]Each sample concentration analyzed 5 times and average is given.
[b]Condition 1 was the use of 1.5 g L^{-1} of $Ca(OH)_2$ nanoparticles at 15 minutes of reaction time.
[c]Condition 2 was the use of 2 g L^{-1} of $Ca(OH)_2$ nanoparticles at 15 minutes of reaction time.
[d]Condition 3 was the use of 1.75 g L^{-1} of CaO nanoparticles at 15 minutes of reaction time.
[e]Condition 4 was the use of 2 g L^{-1} of CaO nanoparticles at 15 minutes of reaction time.

In these, pH and the concentrations of Fe(II), Cu(II), Pb(II), Cd(II), Ni(II), Co(II), and Zn(II) were measured from the supernatant. Sludge volume in each of the selected conditions was also analyzed and used for the leaching study later.

7.1.3.7 Selected treatment conditions study for heavy metals removal

Under all four treatment conditions, the residual concentrations of Pb(II) did not meet the EPA effluent standard (CFR, 40) (0.1 mg L^{-1}). The minimum Pb(II) concentration was found when using 1.5 g L^{-1} of $Ca(OH)_2$ nanoparticles and 1.75 g L^{-1} of CaO nanoparticles. For Cu(II), using 1.5 and 1.75 g L^{-1} of $Ca(OH)_2$ nanoparticles and CaO nanoparticles respectively gave a residual concentration which met the effluent standard limit (2.0 mg L^{-1}). Zn(II), following all selected treatment conditions, had a residual concentration which met its effluent standard limit (5.0 mg L^{-1}); the minimum concentration was obtained when using 1.5 and 1.75 g L^{-1} of $Ca(OH)_2$ and CaO nanoparticles, respectively.

7.1.3.8 Sludge volume study

Table 7.3 shows the sludge volume for all selected treatment conditions. It was found that the final sludge volume depended on the dosage used to precipitate heavy metals. As is shown in Table 7.3, treatment by CaO nanoparticles resulted in lower sludge production. The amount recorded was 68–88 mL L^{-1} for CaO nanoparticles (depending on the dosage), compared to 72–100 mL L^{-1} for $Ca(OH)_2$ nanoparticles, showing that with CaO nanoparticles, the treated wastewater could be more easily filtered. This phenomenon may be attributed to the destructive adsorption effect of CaO nanoparticles (Wagner *et al.* 2000; Oladoja *et al.* 2011). As expected, sludge compression increased with time. Removal of heavy metals using $Ca(OH)_2$ nanoparticles generated a denser sludge than CaO nanoparticles because of its natural property to facilitate precipitation and settling in poorly soluble matter better than those of CaO nanoparticles. Since CaO nanoparticles dissolved slowly compared to $Ca(OH)_2$ nanoparticles, they yielded a low sludge density and volume. For these reasons, the removal of heavy metals would more likely to occur as metal adsorption to the denser sludge. CaO nanoparticles act as seed for nucleation and accumulative growth of metal hydroxides on the surface.

7.1.3.9 Precipitate characterization (XRD)

Precipitates were collected and characterized by XRD. The XRD spectrum (Figure 7.6a and 7.6b) of precipitates shows heavy metals were precipitated as metal hydroxide, and sulfate was removed by precipitation of gypsum ($Ca(SO_4)_2$). XRD spectrum in Figure 7.6a shows a large amount of metals was precipitated along with sulfate (SO_4^{2-}) that was absent in the Figure 7.6b. CaO nanoparticles remove heavy metals but it seems to have less ability to remove sulfate from the sample wastewater.

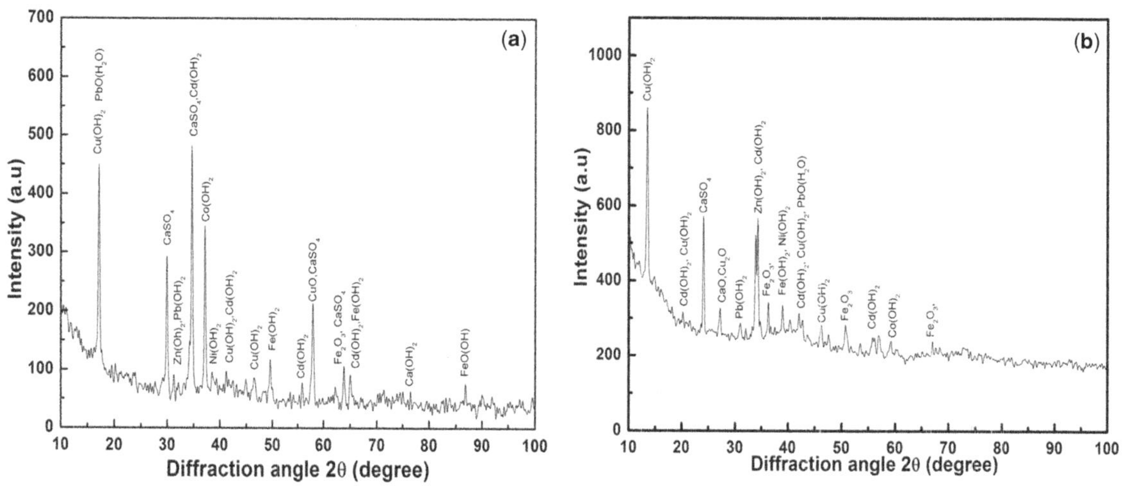

Figure 7.6 The XRD spectrum of precipitate (a) treated with $Ca(OH)_2$ nanoparticles and (b) treated with CaO nanoparticles.

7.1.3.10 Leaching study

Though treatment with Ca-based nanoparticles generates a small amount of sludge, the leaching capacity of heavy metals from the sludge/sediment under specific treatment condition was evaluated. The results obtained from the leaching of sludge are also presented in Table 7.4. Without any pre-treatment sludge/sediment was tested for leaching.

Table 7.4 shows at different conditions small amounts of heavy metals were leached from the sediment or sludge. The heavy metals of interest in the leaching solution were Fe(II), Cu(II), Cd(II), Pb(II), Ni(II), Co(II), and Zn(II). However, of those methods, only Cd(II) and Pb(II) are regulated under the EPA limits. Other regulations were used to compare the Fe(II), Cu(II), Ni(II), Co(II), and Zn(II) concentrations. Leaching of all the metals analyzed was below U.S. Resource Conservation and Recovery Act (RCRA) (CFR, 2003) and California Code of Regulations (CCR, 1985) limits for hazardous materials. Comparing the results with the regulatory criteria, it was found that the leaching of heavy metals from nanoparticle generated sludge/sediment were lower than the regulatory limits. It also can be seen that the release of heavy metals from the CaO nanoparticles precipitates was much lower than that from the $Ca(OH)_2$ nanoparticle precipitates. From the leaching study, one can conclude that the generated sludge/sediment from all selected treatment conditions is fit for inert waste landfill site with relative safety, especially in relation to the contamination of the groundwater.

Table 7.4 The results of heavy metals concentration after leaching test.

Heavy metals	Leachate concentration[a] ($mg\ L^{-1}$) after test				Regulatory limit[b] (Maximum) ($mg\ L^{-1}$)
	Condition 1[c]	Condition 2[d]	Condition 3[e]	Condition 4[f]	
Fe(II)	1.89	2.13	0.88	1.46	–
Cu(II)	1.42	2.65	0.76	1.06	20
Pb(II)	0.42	0.55	0.28	0.39	5.0
Cd(II)	0.41	0.59	0.32	0.44	1.0
Ni(II)	0.91	1.02	0.57	0.98	25
Co(II)	1.05	1.15	0.79	1.02	80
Zn(II)	2.15	4.75	1.52	2.51	250

[a]Each sample concentration analyzed 5 times and average is given.
[b]CFR limits for Pb(II), Cd(II); CCR limits for Zn(II), Ni(II), Co(II), and Cu(II).
[c]Condition 1 was the use of 1.5 g L^{-1} of $Ca(OH)_2$ nanoparticles at 15 minutes of reaction time.
[d]Condition 2 was the use of 2 g L^{-1} of $Ca(OH)_2$ nanoparticles at 15 minutes of reaction time.
[e]Condition 3 was the use of 1.75 g L^{-1} of CaO nanoparticles at 15 minutes of reaction time.
[f]Condition 4 was the use of 2 g L^{-1} of CaO nanoparticles at 15 minutes of reaction time.

7.1.3.11 Dose dependent study for heavy metals removal using CaS nanoparticles

The removal of heavy metals from synthetic effluent was investigated using $Ca(OH)_2$ and CaO nanoparticles and lime at nine dosage levels (0.1, 0.2, 0.4, 0.6, 0.8, 1.0, 1.2, 1.4, 1.6, 1.8 and 2.0 ($\pm$0.01) g L^{-1}).

From Figure 7.7, it is observed that as the bulk CaS dose increased, the heavy metals removal efficiency also increased. For the case of Cu(II), the removal efficiency reached a maximum of 99.95% using a dose of 1.8 g L^{-1}. It is also observed that Fe(II) removal efficiency reached a maximum of 90.02%. In the case of other heavy metals, removal efficiencies $\sim$ 99.56% were achieved using 1.8 g L^{-1} and if the dose was increased, there was no significant change and it implies that at a dose of 1.8 g L^{-1}, the heavy metals removal reaction reached a steady state condition. Dose dependent heavy metals removal efficiency using CaS nanoparticles is shown in Figure 7.8. It was observed that using 1.0 g L^{-1} dose of CaS nanoparticles, nearly 100% removal of Cu(II) was achieved. Other metals like Pb(II), Ni(II), Cd(II), and Co(II) were removed efficiently up to 98, 97, 98, and 97%, respectively, using 1.0 g L^{-1} of CaS nanoparticles. For the case of Fe(II), a 93% removal efficiency was achieved. It is clear that dissolved metal concentrations are lowered dramatically when the molar ratio of sulfide to metal is greater than 1 because of the lower solubility of metal sulfides (Table 4.1) compared to that of metal hydroxides. However, the overdosage of sulfide would result in residual sulfide, which causes a malodorous problem. It is also observed that if dose incrosses, there is no change and it reaches a steady state condition. Comparing the test results it is observed that higher heavy metal removal capability of CaS nanoparticles ia achievable and to reach maximum removal of heavy metal requires doses higher than that using bulk CaS.

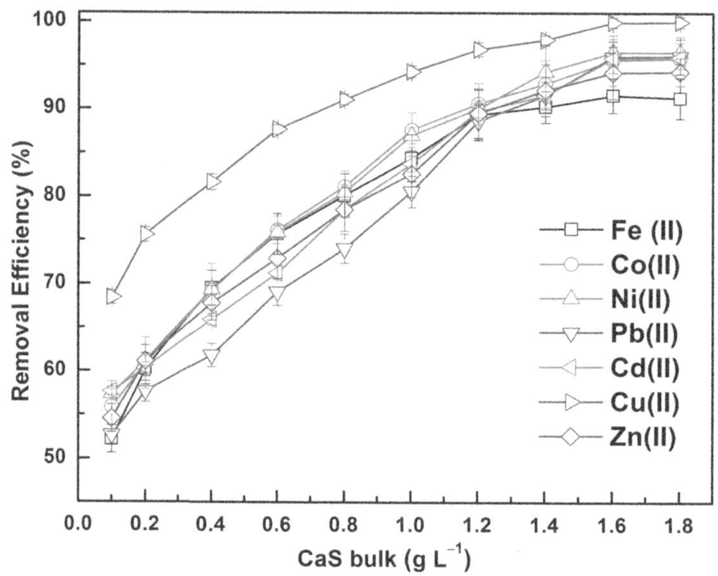

Figure 7.7 Bulk CaS dose dependent heavy metal removals.

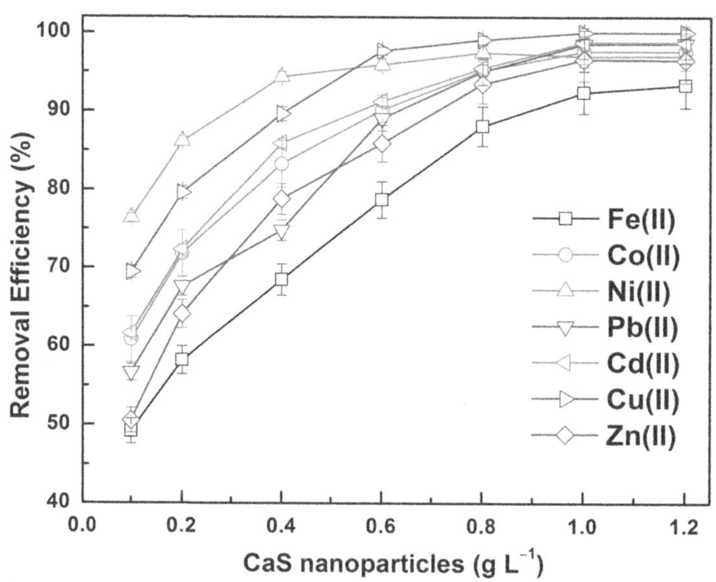

Figure 7.8 CaS nanoparticles dose dependent heavy metal removals.

7.1.3.12 pH dependent study for heavy metals removal using CaS nanoparticles

To study the effect of pH on heavy metals removal from synthetic wastewater using CaS nanoparticles, a pH dependent study was performed using a CaS nanoparticles dose of 1.0 g L^{-1} for various pH conditions. The thermodynamic equilibria involved in metal sulfide precipitation can be expressed as:

$$H_2S \overset{K_{p1}}{\rightleftharpoons} HS^- + H^+, K_{p1} = \frac{[HS^-][H^+]}{[H_2S]}, pK_1 = 6.99 \quad (7.1.8)$$

$$HS^- \overset{K_{p2}}{\rightleftharpoons} S^{2-} + H^+, K_{p2} = \frac{[S^{2-}][H^+]}{[HS^-]}, pK_2 = 17.4 \quad (7.1.9)$$

$$M^{2+} + HS^- \rightleftharpoons MS(s) \quad (7.1.10)$$

$$M^{2+} + S^{2-} \rightleftharpoons MS(s) + H^+ \quad (7.1.11)$$

The concentration of sulfur species is a strong function of pH, as shown in Figure 7.9. The pK_2 value in equation 7.1.9 is currently the most reliable value, as measured by Migdisov *et al.* (2002). From Figure 7.9, it is observed that above pH 7.0, HS^- species predominant in the aqueous environment. At pH 7.0, the bisulfide ion and aqueous H_2S, in solution, are in equal proportions. At pH 7.0 and above, it is suitable for metal sulfide precipitation (Lewis, 2010; Fu & Wang, 2011). Figure 7.10 shows concentration ratio of $[C]/[C_0]$ different heavy metals at different pH conditions. From Figure 7.10, it is observed that as the pH increased, heavy metals removal efficiency also increased. At lower pH, most of the CaS nanoparticles started to dissolve and produced H_2S gas which escaped from aurous solution without reacting with heavy metals. Thus, at low pH less heavy metals were removed. At pH 7.0, most of the metals were precipitated as metal sulfide except Co(II) and Fe(II). As the pH of the wastewater increased to 8.5, all metals were removed and the heavy metals removal efficiency reached up to 98.05–99.02%. It was observed that to

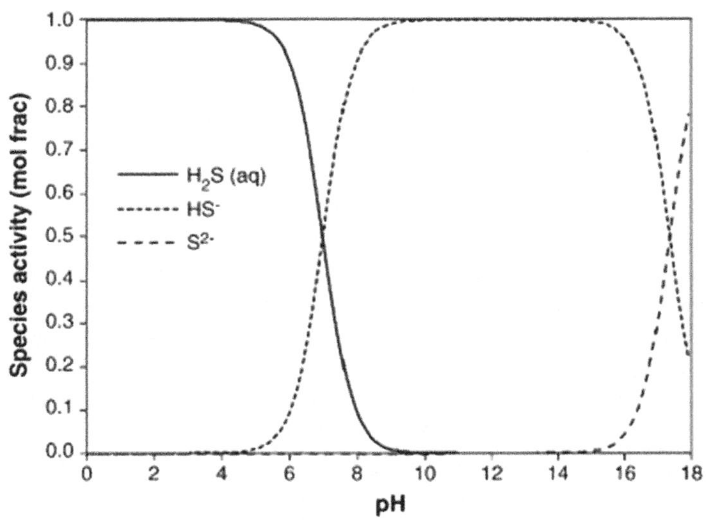

Figure 7.9 The pH dependence of sulphide speciation using MINTEQA2 software.

treat heavy metals using CaS nanoparticles, the pH should be in the range of 7.0–8.0 from mixed heavy metals-containing wastewater.

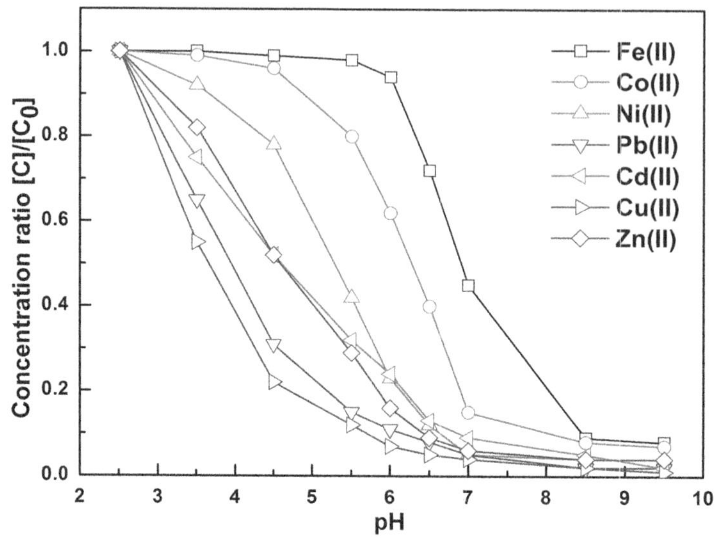

Figure 7.10 The pH dependent heavy metal removal from synthetic wastewater using CaS nanoparticles.

7.2 PERFORMANCE OF γ-Fe_2O_3 NANOTUBES IN HEAVY METALS REMOVAL

7.2.1 Introduction

Magnetic adsorption and separation can be one of the more promising ways for a novel environmental purification technique due to producing little or no flocculants and having the capability of treating large amounts of wastewater within a short time. Moreover, this approach can particularly be adopted when the problem of separation is complex, that is, when polluted water contains solid residues which exclude their treatment in column with regards to the risks of choking.

Only limited research on its application in the environmental area were reported. While magnetite (Hu *et al.* 2004) and maghemite (Hu *et al.* 2005) nanoparticles have been applied to the removal of Cr(VI), chitosan-bound Fe_3O_4 magnetic nanoparticles were prepared for the removal of Cu(II) ions (Chang & Chen, 2005). Ngomsik *et al.* (2005) provided a mini review on the application of magnetic nano and micro particles in the removal of metal ions in wastewaters. Recently, mono-disperse Fe_3O_4 nanocrystals were exploited to remove arsenic from water with magnetic separations at very low magnetic field gradients (Yavuz *et al.* 2006a). By reducing the diameter of Fe_3O_4 nanocrystals from 300 to 12 nm, the removal efficiency of As(III) and As(V) increased by orders of magnitude. Self-assembled 3D flowerlike iron oxide nanostructure materials were shown to have an excellent ability for the removal of As(V), Cr(VI), and Orange II from water (Zhong *et al.* 2006). More recently, superparamagnetic Fe_3O_4 nanoparticles with a surface functionalization of dimercaptosuccinic acid were developed for the removal of toxic soft metal ions such as Hg(II), Ag(II), Pb(II) and Cd(II) (Yantasee *et al.* 2007). Many iron oxide particles with zero-, one-, two- and three-dimensional (0D, 1D, 2D and 3D) nanostructures have been synthesized (Zeng *et al.* 2002;

Zhang *et al.* 2005; Wu *et al.* 2005b; Sun *et al.* 2005; Cesar *et al.* 2006; Wu *et al.* 2007). Ferromagnetic nanotubes were considered as candidates for recording head, biomagnetic sensors, catalysts, and so on, because of their expected vortex magnetization state and floatability in liquid as a result of their hollow structure (Goldstein *et al.* 2001; Haberzettl, 2002; Khizroev *et al.* 2002; Sui *et al.* 2004).

In this section, self-assembled, synthesized in earlier Chapter 4, magnetic adsorbent, maghemite (γ-Fe_2O_3) nanotubes were used to remove heavy metal ions [Cu(II), Cd(II), Co(II), Ni(II), Pb(II), and Zn(II)] from wastewater. Adsorption experiments were conducted systematically in batch experiments to investigate the influence of various factors, such as contact time, initial concentration of metal ions, and pH of the solutions. The applicability of maghemite nanotubes for heavy metal ions removal from water was evaluated in view of the adsorption kinetics, isotherms, and capacity in synthetic mixed metal solution. Reusability of γ-Fe_2O_3 as magnetic adsorbent was also investigated.

7.2.2 Experimental details

7.2.2.1 Materials

Copper sulfate pentahydrate ($CuSO_4{\cdot}5H_2O$), cadmium sulfate octahydrate ($CdSO_4{\cdot}8H_2O$), cobalt sulfate heptahydrate ($CoSO_4{\cdot}7H_2O$), nickel sulfate heptahydrate ($NiSO_4{\cdot}7H_2O$), lead nitrate ($Pb(NO_3)_2$), and zinc sulfate heptahydrate ($ZnSO_4{\cdot}7H_2O$) were purchased from Marck India Co. and standard samples were purchased from Fisher Scientific, Houston, TX, USA. Synthesized maghemite (γ-Fe_2O_3) nanotubes were used as magnetic adsorbent.

7.2.2.2 Adsorption experiments

The adsorption experiments using Cu(II), Cd(II), Co(II), Ni(II), Pb(II), and Zn(II) were performed in the batch method. The conical flasks containing 0.05 g adsorbent and 100 mL of metal ion solutions with the initial pH value 6.0 ± 0.1, were placed on a constant temperature bath oscillator to vibrate at room temperature ($25 \pm 1°C$). After a period of time (4 hours), the magnetic adsorbent was separated from the solutions using a permanent magnet and the initial and final metal ions concentrations were determined by atomic absorption spectrometer (GBC SensAA).

The amount of metal ions adsorbed onto the magnetic adsorbent was calculated according to Equation 7.1.1.

The equilibrium adsorption capacity, q_e (mg g^{-1}) of aqueous heavy metal ions was calculated using the mass balance Equation 4.1.6.

7.2.2.3 Effect of equilibration time

The effect of contact time for each metal ion adsorption was studied using different time intervals ranging from 5 minutes to 240 minutes with an initial metal ion concentration of 100 mg L^{-1}. After the experiment was completed, conical flasks were taken out and the magnetic adsorbents were separated followed by the determination of the residual metal ion concentration.

7.2.2.4 Effect of initial metal ions concentration

Adsorption isotherms were obtained by equilibrating the magnetic adsorbent with metal ion solutions of different initial concentrations ranging from 20 to 400 mg L^{-1} for 4 hours contact time. After separation, the final concentration of heavy metal ions in solution were measured. The adsorption kinetics and isotherm experiments were conducted in duplicate and the average of the two values was used in the calculations.

The maximum difference between the two values was less than 3% of the average value. Obtained results were modelled using Langmuir and Freundlich adsorption equations.

7.2.2.5 Effect of pH

The adsorption behavior of Cu(II), Cd(II), Co(II), Ni(II), Pb(II), and Zn(II) for the same initial concentration and equilibration time was studied as a function of pH. The initial pH values were adjusted from 2.0 to 7.0, using 0.1 M HCl or NaOH solution. After contacting for 4 hours, the suspensions were separated and the residual metal ions concentrations were analyzed.

7.2.2.6 Desorption experiments

The desorption studies of Cu(II), Cd(II), Co(II), Ni(II), Pb(II), and Zn(II) from maghemite nanotubes was performed by using HNO_3. After determination of metal content of the final solution, the adsorbent was washed with excess distilled water in order to reuse in the next experiment. Consecutive sorption-desorption cycles were repeated six times to establish the reusability of the adsorbent.

7.2.3 Results and discussion

7.2.3.1 Adsorption kinetics

The results of adsorption studies canducted as a function of contact time involving Cu(II), Cd(II), Co(II), Ni(II), Pb(II), and Zn(II), are presented in Figure 7.11.

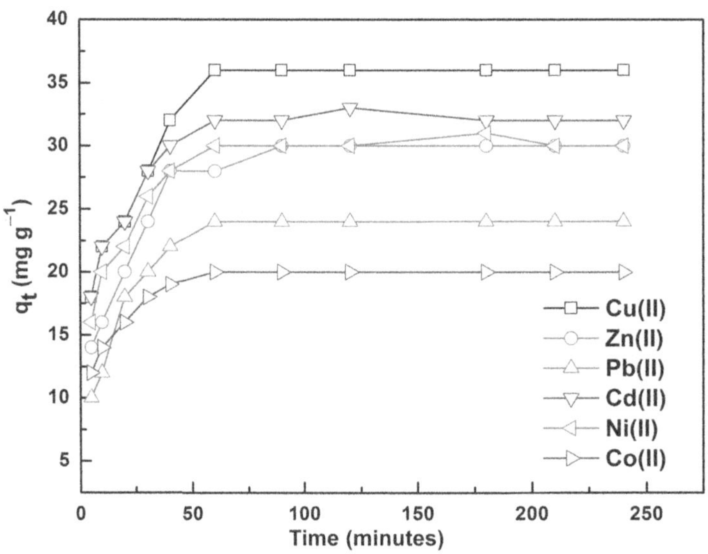

Figure 7.11 Adsorption kinetic curves of heavy metal ions by maghemite nanotubes. Reprinted from 'Removal of Cu(II), Zn(II) and Pb(II) from water using microwave-assisted synthesized maghemite nanotubes' by A. Roy, and J. Bhattacharya, 2012, Chemical Engineering Journal, Pages No. 493–500. Copyright @ 2012 by Elsevier. Reprinted with permission from Elsevier.

The behavior suggested that the removal of Cu(II), Cd(II), Co(II), Ni(II), Pb(II), and Zn(II) by the maghemite nanotube adsorbents occurred in two distinct steps: a relatively rapid phase (first 40 minutes), followed by a increase until the equilibrium was reached. The necessary time to reach the equilibrium was about 60 minutes. Though there was a slight increase of adsorption quantity for Cd(II), Ni(II), and Cu(II) after 120, 180, and 200 minutes, it did not result in any notable different effect. So a contact time of 240 minutes was chosen for further experiments.

In order to determine the rate constants, pseudo-first order and pseudo-second order kinetic were employed to model the adsorption data over the entire time range. The pseudo-first order equation of Lagergren (1898) is expressed as follows:

$$\frac{dp}{dt} = k_1(q_e - q_t) \tag{7.2.1}$$

where q_e and q_t are the amount of metal ions adsorbed per unit weight of sorbent at equilibrium and at time t, respectively (mg g^{-1} dry adsorbent), and k_1 is the rate constant of pseudo-first order adsorption ($min.^{-1}$). Given the boundary conditions of $t = 0$, $q = 0$, Equation 7.2.1 can be integrated to give (Ho & McKay, 1999):

$$\log(q_e - q_t) = \log q_e - \left(\frac{k_1}{2.302}\right) t \tag{7.2.2}$$

If the adsorption process can be described by a pseudo-first order equation, there should be a linear relationship between $\log(q_e - q_t)$ and t. The pseudo-second order equation assumes that the adsorption process involves a chemisorption mechanism and the rate of site occupation is proportional to the square of the number of unoccupied sites. If the rate of adsorption is a second order mechanism, the pseudo-second order chemisorption kinetic rate equation can be expressed as (Ho & McKay, 1999)

$$\frac{dp}{dt} = k_2(q_e - q_t)^2 \tag{7.2.3}$$

where k_2 is a pseudo-second order rate constant (g mg^{-1} min^{-1}). After integrating and applying the boundary conditions for $t = 0$, $q = 0$, Equation 7.2.3 becomes:

$$\frac{t}{q_t} = \frac{1}{(k_2 q_e^2)} + \frac{t}{q_e} \tag{7.2.4}$$

The rate constant k_2 can be obtained from the intercept of the graphed linearized pseudo-second order rate equation. If the pseudo-second order equation can fit the adsorption data, there should be a linear relationship between t/q_t and t.

The relationship between $\log(q_e - q_t)$ versus time t in this work for all three heavy metal ions was not linear over the entire time range, indicating that more than one mechanism was involved in the adsorption, whereas the t/q_t versus t relationship was well fitted in a linear manner with the experimental data. Figure 7.12, shows that the pseudo-second order equation was applicable to all (Cu(II), Cd(II), Co(II), Ni(II), Pb(II), and Zn(II)) adsorption data ($R^2 = 0.999$, $p < 0.0001$). The q_e and k_2 values obtained for Cu(II), Cd(II), Co(II), Ni(II), Pb(II), and Zn(II) by linear regression of t/q against t were 37.31, 32.67, 6.70, 6.80, 24.87, and 31.25 mg g^{-1} adsorbent and 0.0041, 0.0074, 0.016, 0.0071, 0.0065, and 0.0044 g

mg^{-1} min^{-1} respectively. The selectivity of the adsorption capacity of maghemite nanotubes for selected heavy metals was in the order of Cu(II) > Cd(II) > Zn(II) > Ni(II) > Co(II) > Pb(II).

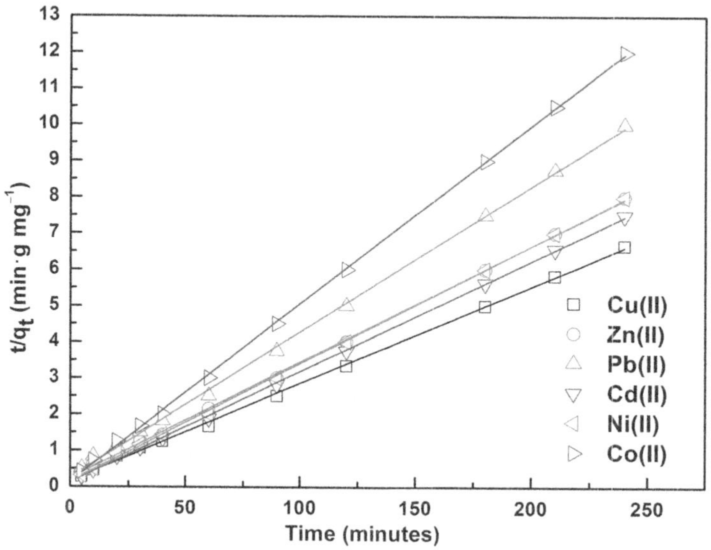

Figure 7.12 Test of pseudo-second order kinetic for heavy metal ions onto maghemite nanotubes. Reprinted from 'Removal of Cu(II), Zn(II) and Pb(II) from water using microwave-assisted synthesized maghemite nanotubes' by A. Roy, and J. Bhattacharya, 2012, Chemical Engineering Journal, Pages No. 493–500. Copyright @ 2012 by Elsevier. Reprinted with permission from Elsevier.

7.2.3.2 Effect of initial metal ions concentration with isothermal models

The metal ion removal performance of the maghemite nanotubes was evaluated as a function of the initial metal ions concentration (from 20 mg L^{-1} to 400 mg L^{-1}) at a pH value of 6.0 ± 1. In this study, the adsorption time was fixed at 240 min because the adsorption of Cu(II), Cd(II), Co(II), Ni(II), Pb(II), and Zn(II) ions onto the magnetic nanotubes sufficiently reached an equilibrium state after 240 minutes contact time based on the contact time dependency test.

As illustrated in Figure 7.13, all the metal ions (Cu(II), Cd(II), Co(II), Ni(II), Pb(II), and Zn(II)) adsorption capacity depended on the initial heavy metal ion concentration. Notably when the initial metal ions concentration was lower than ca. 100 mg L^{-1}, the metal ions (Cu(II), Cd(II), Co(II), Ni(II), Pb(II), and Zn(II)) adsorption capacity relatively increased with an increase of the initial metal ions concentration. In the range of these concentrations, Cu(II), Cd(II), Co(II), Ni(II), Pb(II), and Zn(II) ions could bind to the abundant adsorption sites on the surface of the maghemite nanotubes, leading to the distinctively increased adsorptivity of the maghemite nanotubes. Above 100 mg L^{-1} of initial heavy metal ion concentration, the rate of adsorption capacity became gradually slower as the initial metal ions concentration increased. The adsorption capacity of maghemite nanotubes for Cu(II) ions was higher than that exhibited for Zn(II) and followed by Pb(II) ions.

In order to determine the adsorption capacity of maghemite nanotube adsorbents towards the examined heavy metal ions, adsorption studies over a large initial concentration range from 20 to 400 mg L^{-1} were performed. The maximum adsorption capacities of the maghemite nanotube adsorbent for heavy metal ions were evaluated using the adsorption isotherms. Two different isotherm models (Langmuir and Freundlich), were applied to the obtained adsorption data. In the Langmuir model, it is assumed that all the adsorption sites of the adsorbent have an identical binding energy and each site binds to only a single adsorbate (Wan *et al.* 2002). The Langmuir isotherm was discussed in Section 4.1.8.2 and the Freundlich isotherm was discussed in Section 4.1.8.3.

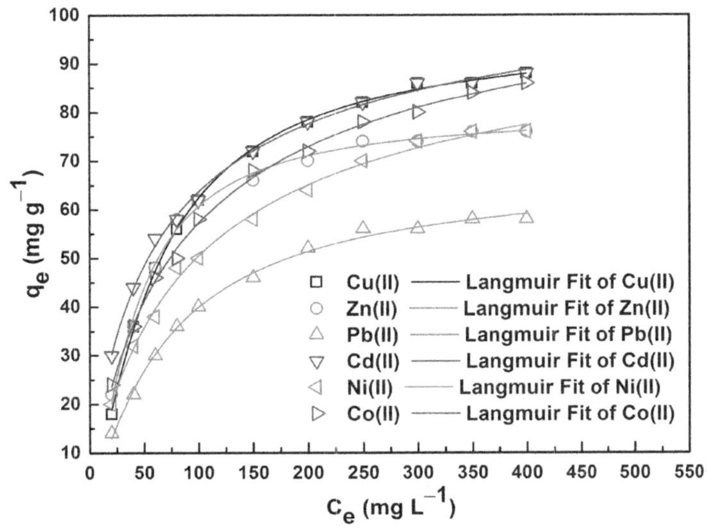

Figure 7.13 Adsorption isotherm of heavy metal ions by maghemite nanotube adsorbents at pH 6.0 ± 0.1 and T 25 ± 1°C. Reprinted from 'Removal of Cu(II), Zn(II) and Pb(II) from water using microwave-assisted synthesized maghemite nanotubes' by A. Roy, and J. Bhattacharya, 2012, Chemical Engineering Journal, Pages No. 493–500. Copyright @ 2012 by Elsevier. Reprinted with permission from Elsevier.

The quantitative relationship between initial Cu(II), Cd(II), Co(II), Ni(II), Pb(II), and Zn(II) ions concentration and the adsorption capacity was analyzed with two different isotherm models (Figure 7.14 and Figure 7.15). The calculated correlation coefficients (b, q_m, n, and K_f) and linear regression coefficient (R^2) values for both the Langmuir and Freundlich model are shown in Table 7.5.

The results showed that the values of correlation coefficient for the adsorption of Cu(II), Cd(II), Co(II), Ni(II), Pb(II), and Zn(II) onto maghemite nanotube adsorbents usinf the Langmuir adsorption model were 0.995, 0.998,0.998, 0.997, 0.997, and 0.997 respectively, which demonstrated the good fitting of experimental data by the Langmuir model. From the Langmuir isotherms, the maximum adsorption capacities of tubular maghemite adsorbents towards Cu(II), Cd(II), Zn(II), Ni(II), Pb(II), and Co(II) were 111.11, 94.33, 86.95, 86.20, 71.42, and 60.60 mg g^{-1}, respectively.

The favorable nature of adsorption can be expressed in terms of a dimensionless separation factor employing an equilibrium parameter, which is defined by Hall *et al.* (1966):

$$R_L = \frac{1}{(1 + bC_0)} \tag{7.2.5}$$

where b is the Langmuir constant and C_0 is the initial concentration of the adsorbate in solution. The values of R_L indicates the type of isotherm to be irreversible ($R_L = 0$), favorable ($0 < R_L < 1$), linear

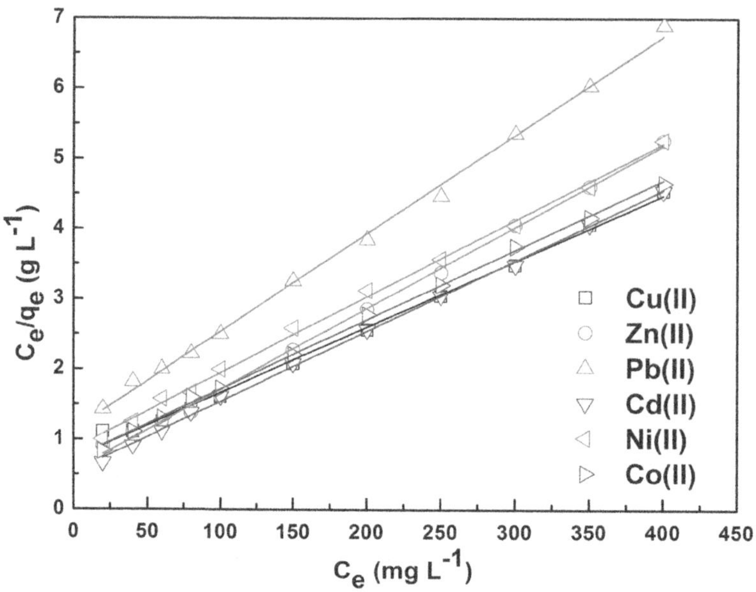

Figure 7.14 Langmuir model fitted test results of adsorption. Reprinted from 'Removal of Cu(II), Zn(II) and Pb(II) from water using microwave-assisted synthesized maghemite nanotubes' by A. Roy, and J. Bhattacharya, 2012, Chemical Engineering Journal, Pages No. 493–500. Copyright @ 2012 by Elsevier. Reprinted with permission from Elsevier.

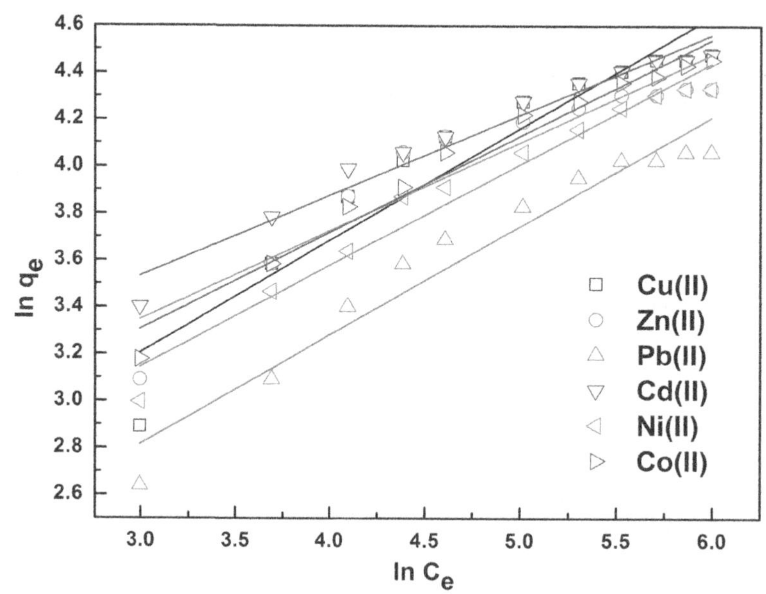

Figure 7.15 Freundlich model fitted test results of adsorption. Reprinted from 'Removal of Cu(II), Zn(II) and Pb(II) from water using microwave-assisted synthesized maghemite nanotubes' by A. Roy, and J. Bhattacharya, 2012, Chemical Engineering Journal, Pages No. 493–500. Copyright @ 2012 by Elsevier. Reprinted with permission from Elsevier.

($R_L = 1$) or unfavorable ($R_L < 1$). The R_L values for Cu(II), Cd(II), Co(II), Ni(II), Pb(II), and Zn(II) onto maghemite nanotube adsorbents were less than 1 and greater than zero indicating favorable adsorption (Table 7.5).

Table 7.5 Adsorption parameters of the Langmuir and Freundlich isotherm models for the adsorption of Cu(II), Cd(II), Co(II), Ni(II), Pb(II), and Zn(II) ions onto maghemite nanotube. Reprinted from 'Removal of Cu(II), Zn(II) and Pb(II) from water using microwave-assisted synthesized maghemite nanotubes' by A. Roy, and J. Bhattacharya, 2012, Chemical Engineering Journal, Pages No. 493–500. Copyright @ 2012 by Elsevier. Reprinted with permission from Elsevier.

Metal ions	Langmuir				Freundlich		
	q_m (mg g^{-1})	b (L mg^{-1})	R^2	R_L[a]	K_f ($mg^{1-(1/n)}L^{1/n}g^{-1}$)	n	R^2
Cu(II)	111.11	0.0123	0.995	0.45	5.87	2.09	0.894
Cd(II)	94.33	0.0228	0.998	0.30	12.24	2.91	0.962
Zn(II)	86.95	0.0207	0.997	0.33	9.17	2.65	0.863
Ni(II)	86.20	0.0014	0.997	0.87	6.36	2.31	0.966
Pb(II)	71.42	0.0123	0.997	0.44	4.13	2.15	0.941
Co(II)	60.60	0.0105	0.998	0.48	7.94	2.42	0.969

[a]R_L measured by using C_0=100 mg L^{-1}

7.2.3.3 Effect of pH

The pH is one of the most important parameters controlling the metal ion adsorption process. The influence of initial pH on the amount adsorbed amount was studied in the range of 2.0 to 7.0. The relationship between the initial pH values and the quantities of heavy metal ions adsorbed on maghemite nanotube adsorbents is presented in Figure 7.16 which showed that the metal ions adsorbed increased as pH increased.

At a low pH adsorption of metal ions is lower. With an increase in pH, the adsorption increases but in different proportions depending on the metal ions, until reaching a plateau. Cu(II) and Cd(II) ions removal rate sharply increases from pH 2.0–5.0, then from pH 5.0–6.0 it reaches an equilibrium state but Pb(II) ions removal rate dropped slightly drop at pH 7.0. For the case of Ni(II) and Co(II) ions, the removal rate drops slightly at pH 3.0 and 7.0, respectively. On the other hand, Zn(II) ions removal rate an increases slowly with increase in pH. In this study, higher initial pH values were not used because at higher pH, all metal ions precipitate from solution in their hydroxide form. Higher pH is favorable for the deprotonation of sorbent surface (Johnson *et al.* 2000). Increased deprotonation results in an increase of the negatively charged sites, which enhances attractive forces between the sorbent surface and the Cu(II), Cd(II), Co(II), Ni(II), Pb(II), and Zn(II) ions, and thus results in an increase in the adsorption capacity. On the other hand, in the lower pH region, the positively charged sites dominate, which enhances the repulsion forces existing between the sorbent surface and the Cu(II), Cd(II), Co(II), Ni(II), Pb(II), and Zn(II) ions, and therefore decreases the adsorption of Cu(II), Cd(II), Co(II), Ni(II), Pb(II), and Zn(II) ions.

Regarding the pH_{PZC} of maghemite nanotubes (pH_{PZC} maghemite nanotube $= 6.8 \pm 1$; Figure 7.17) in solution having pH lower than pH_{PZC} of maghemite nanotube, the surface hydroxyl sites could be protonated and positively charged. On the other hand, the active sites are deprotonated when the solution pH is higher than pH_{PZC}, resulting in negatively charged sites and the adsorption of metal ions on maghemite

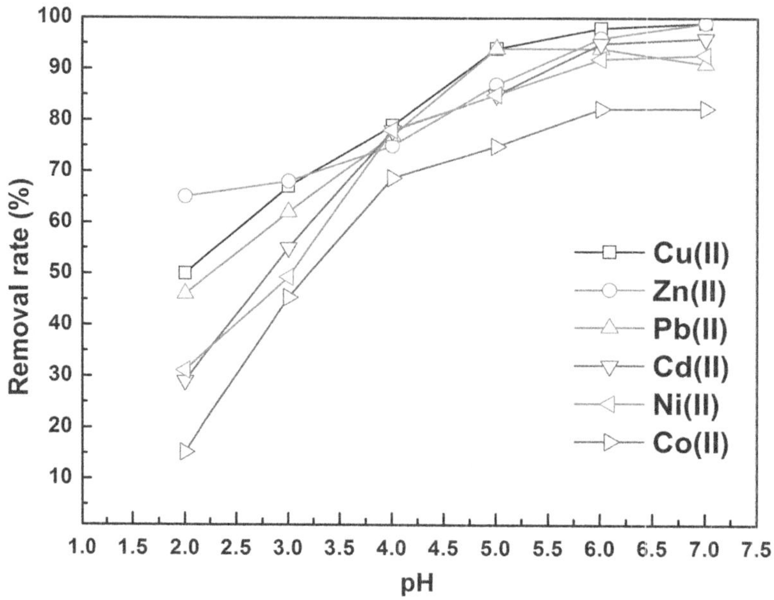

Figure 7.16 Effect of pH on heavy metals removal efficiency by maghemite nanotube. Reprinted from 'Removal of Cu(II), Zn(II) and Pb(II) from water using microwave-assisted synthesized maghemite nanotubes' by A. Roy, and J. Bhattacharya, 2012, Chemical Engineering Journal, Pages No. 493–500. Copyright @ 2012 by Elsevier. Reprinted with permission from Elsevier.

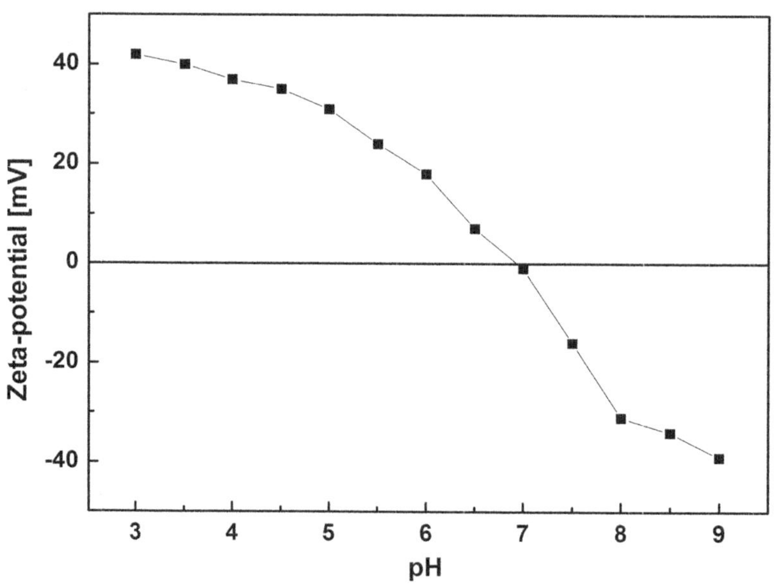

Figure 7.17 Zeta potential of synthesized maghemite nanotubes. Reprinted from 'Removal of Cu(II), Zn(II) and Pb(II) from water using microwave-assisted synthesized maghemite nanotubes' by A. Roy, and J. Bhattacharya, 2012, Chemical Engineering Journal, Pages No. 493–500. Copyright @ 2012 by Elsevier. Reprinted with permission from Elsevier.

nanotubes could take place via electrostatic interaction with the negatively charged sites on maghemite nanotube surface. However, in this study, the removal of metal ions was observed in solutions having pH values less than pH_{PZC} of maghemite nanotube, therefore, the adsorption of metal ions on maghemite nanotube surface would occur via non-electrostatic interaction.

7.2.3.4 Desorption and reusability

In order to make the adsorption process more economical and feasible, the desorption efficiency and regeneration potential of γ-Fe_2O_3 nanotubes were studied. The desorption capacity of γ-Fe_2O_3 nanotubes was calculated according to Equation 4.1.11 stated earlier. From the pH study, it has been found that the adsorption of Cu(II), Cd(II), Co(II), Ni(II), Pb(II), and Zn(II) ions by γ-Fe_2O_3 nanotubes at pH $\leq$ 2.0 was negligible. It indicated that the desorption of metal ions from adsorbent was possible around pH $\leq$ 2.0.

HNO_3 solutions of different pH (2.5, 2.0 and 1.5) were used to examine the desorption study. It was found that desorption percentages increases as pH decreases. The highest desorption was achieved (94, 87, 85, 84, 75 and 72%) in the HNO_3 solutions of pH 1.5 for the Cu(II), Cd(II), Zn(II), Ni(II), Co(II), and Pb(II) ions, respectively. The higher desorption efficiency at lower pH values could be due to the sufficiently high hydrogen ion concentration, which led to the strong competitive adsorption. The reusability of γ-Fe_2O_3 nanotubes as adsorbent was also studied after the desorption process. The reusability of γ-Fe_2O_3 nanotubes, adsorption-desorption cycles were repeated six times using the same adsorbents. Figure 7.18 shows that the adsorption capacities of Cu(II), Cd(II), Co(II), Ni(II), Pb(II), and Zn(II) slightly decreased with the increase in adsorption-desorption cycle; at the end of the sixth cycle, more than 70, 66, 68, 59, 46, and 42% of the initial adsorption capacities were obtained for Cu(II), Cd(II), Zn(II), Ni(II), Co(II), and Pb(II) respectively. The regeneration ability of γ-Fe_2O_3 nanotubes would decrease the operation cost and indicates good applicability.

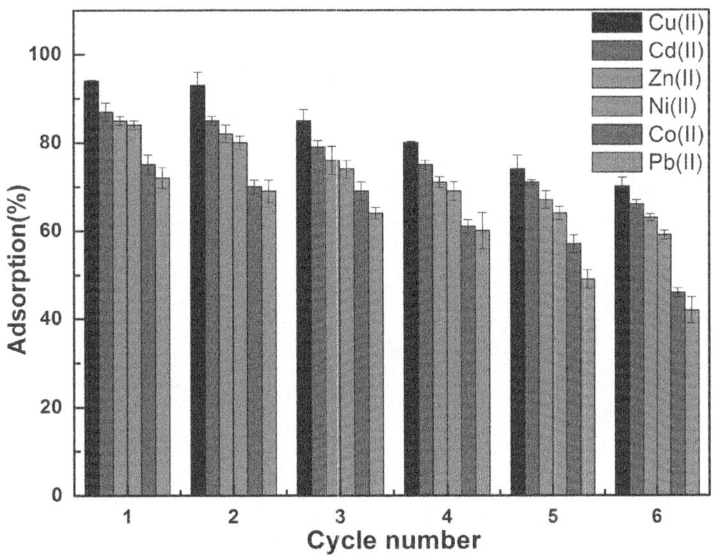

Figure 7.18 Adsorption isotherm of heavy metal ions by maghemite nanotube adsorbents at pH 6.0 $\pm$ 0.1 and T = 25 $\pm$ 1°C. Reprinted from 'A binary and ternary adsorption study of wastewater Cd(II), Ni(II) and Co(II) by γ–Fe_2O_3 nanotubes' by A. Roy, and J. Bhattacharya, 2013, Separation and Purification Technology, Pages No. 172–179. Copyright @ 2013 by Elsevier. Reprinted with permission from Elsevier.

Chapter 8

Continuous treatment of heavy metals by nanomaterials

8.1 INTRODUCTION

This Chapter, discusses the experiments required for multistage continuous bench scale system using different nanomaterials. Synthesized $Ca(OH)_2$ nanoparticles for use as a hydroxide precipitating agent is to be tested in the 1st stage of wastewater treatment. For sulfide precipitation use of CaS nanoparticles is required in the 2nd stage of treatment where 1st stage treated water is used as influent. Magnetic adsorption and filtration is required in final stage of continuous treatment mode using γ-Fe_2O_3.

8.2 EXPERIMENTAL DETAILS

8.2.1 Materials

$Ca(OH)_2$, CaS and γ-Fe_2O_3 nanoparticles were prepared according to Chapter 6. A synthetic heavy metals rich wastewater, containing Fe(II), Cu(II), Cd(II), Co(II), Ni(II), Pb(II), and Zn(II) was prepared from their respective standard reagent grade metal sulfate and nitrate salts (Merck Germany).

Each metal salt, ferrous sulfate heptahydrate ($FeSO_4 \cdot 7H_2O$), copper sulfate pentahydrate ($CuSO_4 \cdot 5H_2O$), cadmium sulfate octahydrate ($CdSO_4 \cdot 8H_2O$), cobalt sulfate heptahydrate ($CoSO_4 \cdot 7H_2O$), nickel sulfate heptahydrate ($NiSO_4 \cdot 7H_2O$), lead nitrate ($Pb(NO_3)_2$), and zinc sulfate heptahydrate ($ZnSO_4 \cdot 7H_2O$) was dissolved in Milli-Q water at a concentration of each heavy metals presented in Table 8.1 for continuous studies. The synthetic wastewater pH was adjusted to 1.99–2.12 by the addition of 1 M H_2SO_4. The composition of the synthetic wastewater was selected because it is within the range of relevant concentrations for real effluents discharged from different industries. Table 6.1 also lists the EPA discharge limit (CFR, 40) of selected heavy metals and the USEPA limit for drinking water (USEPA, 2011).

8.2.2 Bench scale system set-up for the multistage treatment system

To build a laboratory-bench-scale set up for continuous, studies acrylic glass materials was used. Figure 8.1 shows a schematic diagram of the multistage treatment unit.

Table 8.2 lists the properties of the treatment system, consisting of a reservoir (60 cm in length and 25 cm inner diameter, containing 30 L of wastewater), a plug flow reactor (184 cm in length and 5 cm

Table 8.1 Chemical composition of synthetic wastewater and limits for wastewater discharge and drinking water.

Parameter of simulated wastewater		EPA industrial discharge limit (mg L^{-1})	Maximum acceptable limit for drinking (mg L^{-1}) (USEPA, 2011)
pH	1.99–2.12	6.0–9.5	6.8–7.8
Heavy metals concentration (mg L^{-1})			
Fe (II)	750.00	3.0	0.3
Cu (II)	500.00	3.0	1.3
Pb (II)	500.00	0.1	0.015
Cd (II)	500.00	2.0	0.003
Ni (II)	500.00	3.0	0.01
Co (II)	500.00	–	0.01
Zn (II)	500.00	5.0	5.0

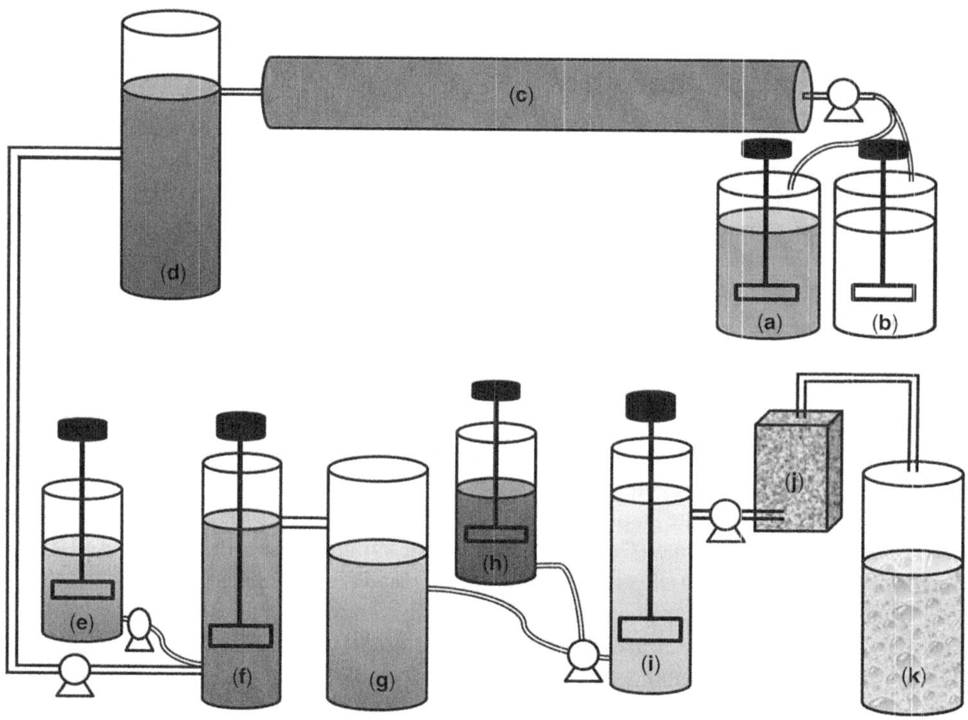

Figure 8.1 Schematic representation of bench scale multistage treatment system. Stock wastewater (a), nanoparticles dispersion tank (b, e, and h), plug flow reactor (c), stirred tank (f and i), settling tank (d and g), magnetic filter (j), and effluent collector (k).

inner diameter, capacity 3.6 L volume), a settling tank (58 cm in length and 15 cm inner diameter, capacity 10.249 L volume), a stirred tank reactor (20 cm in length and 15 cm inner diameter, capacity 3.5 L), a settling tank (58 cm in length and 15 cm inner diameter, capacity 10.249 L volume), a stirred tank reactor

(20 cm in length and 15 cm inner diameter, capacity 3.5 L), and a magnetic filtration unit (10 cm in length, 4 cm width and 7 cm depth, capacity 0.28 L).

Table 8.2 Property of multistage treatment system.

Composition	Volume (L)	Retention time (minutes)	Flow rate (mL min.$^{-1}$)	Dimension (cm)
Reservoir	30	–	–	$L60 \times D25$
Plug flow reactor	3.6	62	58 (26 + 32)	$H184 \times D5$
Settling tank	10.249	15	–	$H58 \times D15$
Stirred tank reactor	3.5	73	48 (16 + 32)	$H20 \times D15$
Settling tank	10.249	15	–	$H58 \times D15$
Stirred tank reactor	3.5	73	48 (24 + 24)	$H20 \times D15$
Magnetic filter unit	0.28	8.75	32	$H10 \times W4 \times D7$

The nanoparticles were collected in a magnetic filter which is composed of a cartridge (V = 280 cm^3) filled with stainless steel wool (DIN 1.4460, fiber: length $\approx$ 20 cm, diameter = 0.06 mm, magnetic saturation: 80 Am^{2}g^{-1}). The magnetic field is induced by a permanent magnet, which surrounded the cartridge.

8.2.3 Operation of the multistage treatment unit

In the first stage, $Ca(OH)_2$ nanoparticles were dispersed in Milli Q (Millipore Synergy) water and a synthetic heavy metals rich water (pH = 2.45 ± 0.02) was kept in a reservoir. The dissolved nanoparticles tank and reservoir were continuously stirred to avoid agglomeration using high speed mechanical direct drive stirrers (Remi laboratory instruments). Using a Miclins multichannel peristaltic pump (pp–20–4c–EX), flow rates of nanoparticles solution and synthetic heavy metal water were controlled list range of flow rates. Before entering into the plug flow reactor (c) two silicone tubes carrying two separate solutions were connected by a Y-connector. The flow rate of the influent solutions was the combined flow rate (58.72 mL min.$^{-1}$) from two silicon tubes. $Ca(OH)_2$ nanoparticles dose was varied from 0.05 to 0.3 g L^{-1} with a constant flow rate and constant temperature (at room temperature, 25 ± 0.1°C). A tracer study was conducted using methylene blue as a tracer with concentration of 10 mg L^{-1} into the reactor using a flow rate of 58.72 mL min.$^{-1}$ (τ = 62 minutes). Samples were collected at appropriate time intervals and analyzed for methylene blue levels in the reactor by measuring the absorbance at 664 nm using a UV–V is spectrophotometer (Shimadzu UV–1800). Tracer study showed the $d = D/uL_l$ (Metcalf *et al.* 2010) (dispersion number) is 0.0022. The flow characteristic of the reactor was classified as an approaching plug flow pattern. Theoretical hydraulic retention time (HRT) was similar to the HRT as calculated from the tracer study. After the first stage of treatment, effluents were allowed to settle in the settling chamber for 30 minutes. After that, the effluents were transported using a peristaltic pump in the second stage of the reactor (f), where well-dispersed CaS nanoparticles were also transported and used as the sulfide precipitating agent. Like the previous reactor, two silicon tubes carrying two solutions were connected using a Y-connector and then entered into the second reactor which was a continuous stirred tank reactor. The inlet flow rate was controlled at 48 mL min.$^{-1}$, the combined flow rate of the flow from the two tubes. The CaS nanoparticles dose was varied from 0.1 to 3 g L^{-1} with a constant flow rate and constant temperature (at room temperature, 25 ± 2°C). As before, the tracer study was performed to evaluate whether complete mixing had taken place in the second reactor or not. The reactor was ideally mixed

CSTR (as confirmed by a retention time distribution test using NaCl as a tracer) with flow rate of 48 mL $min.^{-1}$. The difference between the residence time obtained from the experimental tracer data and the theoretical value ($\tau = V/Q$) varies 2.2% for the 48 mL $min.^{-1}$ flow rate of the experiment. After treatment with CaS nanoparticles, effluent water was allowed to settle down in the settling tank for a period of 30 minutes. In the third reactor (i), well dispersed magnetic nanomaterials (γ-Fe_2O_3) and a second reactor effluent entered using a Y-connector. The combined flow rate from the two tubes was maintained at 48 mL $min.^{-1}$ by a peristaltic pump and a designed HRT of 73 minutes for complete adsorption of heavy metals by magnetic nanomaetrials. The magnetic nanoparticles dose varied from 0.05 to 0.3 g L^{-1}. The reacted solution was pumped out and passed through a magnetic filtration unit (j). The superficial flow velocity was 1.16 mL $min.^{-1}$ and the HRT was 8 minutes. The filtered water was collected and analyzed for heavy metal ions and sulfate ion concentrations. The agitational devices (100–150 rpm) were set up in the CSTRs for homogeneous mixing and the agitational devices (200–250 rpm) were set up in the nanoparticles dispersion tanks.

8.2.4 Analytical method

Samples of both the influent and effluent waters were taken at specific time intervals in each stage of the treatment process for major physic-chemical parameters (pH, electric conductivity E_h relative to a H_2 electrode, and water temperature) using an Orion 5-Star multi-probe bench top operators. For each sample, two amber narrow mouth polypropylene bottles (Tarsons Products Pvt. Ltd.) were filled, one of which was acidified (for total cation and metal analysis) and other was filtered (with 0.45 μm cellulose nitrate filters) prior to acidification (for dissolved cation and metal analyses). All samples were analyzed (within the same day of sampling) for metals using ASS (GBC SensAA). Replicate samples ($n = 5$) were taken periodically, rather than routinely, for logistical reasons, in accordance with Quality Assurance/Quality Control procedures detailed by APHA (APHA, 1988). X-ray Diffraction (XRD) analyses were performed on selected freeze-dried powdered sample precipitate crusts using a Philips X-pert Pro X-ray diffractometer with a Co Kα radiation source (λ = 1.789 Å) at a scan speed of 2 ° $min.^{-1}$. Phase identification was performed by means of an ICDD database. Solid samples were also analyzed using field emission scanning electron microscope (FESEM) of Carl Zeiss model Supra-40 (with an accelerated voltage of 10–20 kV).

8.2.5 Calculations

The treatment efficiency for the target metal ions was obtained using the following equation:

$$\text{Treatment efficiency (\%)} = \frac{C_0 - C_f}{C_0} \times 100 \tag{8.2.1}$$

where C_0 = influent contaminant concentration (mg L^{-1}) and C_f = effluent contaminant concentration (mg L^{-1}).

The rate constant (K) of each heavy metal precipitation reaction was calculated using the following equation for the case of reactor (c):

$$K_c = \frac{[ln(C_0) - ln(C_f)] \times Q}{V} \tag{8.2.2}$$

where C_0 = influent contaminant concentration (mg L^{-1}), C_f = effluent contaminant concentration (mg L^{-1}), V = reactor volume, and Q = flow rate (combined).

Rate constant (K) of each heavy metal precipitation reaction was calculated using the following equation for the case of reactor (f):

$$K_f = \frac{[C_0/C_f - 1] \times Q}{V} \tag{8.2.3}$$

where C_0 = influent contaminant concentration (mg L^{-1}), C_f = effluent contaminant concentration (mg L^{-1}) and V = reactor volume, and Q = flow rate (combined)

The rate constant (K) of each heavy metal adsorption reaction was calculated using pseudo-second order following equation for the case of reactor (i):

$$\frac{t}{q_t} = \frac{1}{(K_i q_e^2)} + \frac{t}{q_e} \tag{8.2.4}$$

where q_e and q_t are the amount of metal ions adsorbed per unit weight of sorbent at equilibrium and at time t, respectively (g mg^{-1} dry adsorbent) and k_i is pseudo-second order rate constant (g mg^{-1} min^{-1}).

8.3 RESULTS AND DISCUSSION

8.3.1 Performance of plug flow reactor (c)

In bench scale multi-stage continuous treatment, synthetic wastewater was treated in three steps. In the first stage, synthetic wastewater was treated with $Ca(OH)_2$ nanoparticles in reactor (c). Here, nanoparticles were dispensed in Mili-Q water and the dose of the nanoparticles were varied from 0.5 to 3.0 g L^{-1} and other parameters such as flow rate and heavy metal load (3750 mg L^{-1}), were kept constant. The performance of the plug flow reactor throughout the study is indicated in Figure 8.2. It was observed that as the nanoparticles dose was increased, the heavy metal removal efficiency also increased. At the initial dose of $Ca(OH)_2$ nanoparticles (0.5 g L^{-1}), (Figure 6.2a) Fe(II) was removed about 60%, except Zn(II), all metal were removed by 47–50%, and the effluent pH increased from 1.99 to 4.0. As the dose increased to 1.0 g L^{-1} (Figure 8.2b), it was observed that the Cu(II) removal reached maximum of 80–82%. For the case of other metals their removal rate also increased. The pH of the effluents after the dose treatment was observed ranging from 5.0–5.5. Higher removals of Cu(II) suggested that at such pH, Cu(II) starts to precipitate as copper hydroxide ($Cu(OH)_2$) (Ahmad & Afzal, 2001). Figure 8.2c showed removal of all heavy metals increased as the effluent pH was increased to 7.0–7.5. Metal oxides, hydroxides, and oxy-hydroxides, especially iron hydroxide can scavenge other metals such as Cu(II), Ni(II), Cd(II), Co(II), and Zn(II), from the free water column in a process called co-precipitation (Ahmad & Afzal, 2001; Nkegbe, 2005). During co-precipitation, trace metal cations adsorb to negative charged surface sites of metal hydroxide and oxy-hydroxide precipitates (Al-llil, 2010). This phenomenon usually occurs on metal hydroxides and oxy-hydroxides; co-precipitation primarily occurs under aerobic condition. The XRD study (Figure 7.5a) of the precipitate confirmed co-precipitation also occurred in the experiment. The steady state heavy metal removal efficiency reached a maximum with a dose of 2.5 g L^{-1}.

At this dose all heavy metals except Zn(II) had a removal efficiency of 97–99%. One reason may be due to the amphoteric nature of zinc hydroxide, that with the increase of pH starts to redissolve. However, the effluent pH was found quite high compared to the EPA (CFR, 40) discharge limit. According to the EPA, the pH of the effluent should be in the range of 6.0–9.5 and for drinking water purposes it should be in the

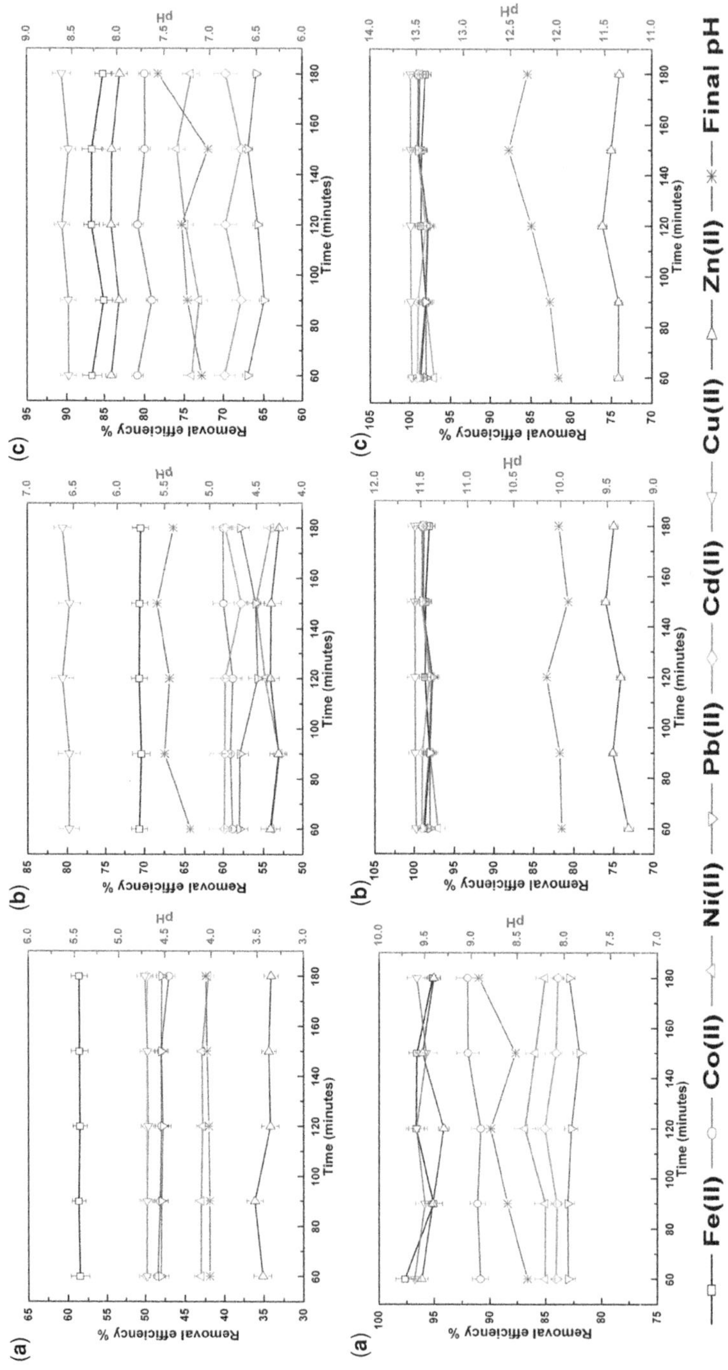

Figure 8.2 Dose dependent continuous study using $Ca(OH)_2$ nanoparticles, 0.5 (a), 1.0 (b), 1.5 (c), 2.0 (d), 2.5 (e), and 3.0 (f) $g\ L^{-1}$.

range of 6.8–7.8. Using 1.5 and 1.0 g L^{-1} dosages, the heavy metal removal efficiency obtained was 82–97% and 65–90%, respectively, with pH of the effluents being 8.5–9.0 and 7.0–7.5. Metal hydroxides are amphoteric, i.e., they are increasingly soluble at both low and high pH, and the point of minimum solubility (optimum pH of precipitation) occurs at a different pH value for every metal. At a pH at which the solubility of one metal hydroxide may be minimized, the solubility of another may be relatively high. Since metal hydroxides are quite soluble, many such hydroxides will start to go back into solution if the pH changes slightly. The rate constant of the heavy metal removal reaction was calculated using Equation 8.2.2 and it was observed that it varies with the dose of nanoparticles. At dose 1.0 and 2.5 g L^{-1}, the rate constant (k_c) of Fe(II), Co(II), Ni(II), Pb(II)), Cd(II), Cu(II), and Zn(II) were 0.020, 0.015, 0.013, 0.014, 0.015, 0.027, 0.013 and 0.070, 0.073, 0.068, 0.070, 0.076, 0.111, 0.00008 min^{-1}, respectively. Images of the continuous treatment of wastewater using the reactor (c) was provided in Figure B.1 in **Appendix** B.

8.3.2 Performance of stirred tank reactor (f)

Literature (Lewis, 2010; Fu & Wang, 2011) suggested treating heavy metals using sulphidic precipitating agents such as CaS, Na_2S, etc. The pH of the influent should be above 7.0. At this pH range, HS^- predominates which helps in heavy metal precipitation as the metal sulfide (Figure 7.9). Batch tests were performed to evaluate the effective pH to remove heavy metals from the wastewater using CaS nanoparticles. It was observed that, at pH 7.0, all heavy metals were removed (Figure 7.10) except for Fe(II) and Co(II). Therefore, the effluent of $Ca(OH)_2$ nanoparticles was treated at a dose of 1.0 g L^{-1}, and was used as the influent for this reactor as pH ranged from 7.0–7.5. Under these conditions, the rapid generation of supersaturation is controlled by the rate of dissolution of the hydrogen sulfide (H_2S) gas (Ter Maat *et al.* 2005). The influent was treated with CaS nanoparticles using doses ranging from 0.1–2.5 g L^{-1}. Figure 8.3 shows the performance of heavy metal removal using CaS nanoparticles; other parameters such as flow rate and HRT were kept constant. With the increase in dose, the heavy metal removal efficiency also increased. Figure 8.3a showed all metals were removed 90–95%, except for Co(II) which was removed up to 76–77%. With the dose of 0.5 g L^{-1}, Cu(II) removal exceeded 98%; other heavy metals removal ranged from 92–96%, and Co(II) removal was 86%. As dose was increased in the continuous treatment, the heavy metals removal efficiency also increased. These results indicated that as the dose of CaS nanoparticles increased, the total $[HS^-]$ to total $[Me^{2+}]$ ratio reached 1.0. During sulfide precipitation, it was observed from Figure 8.3 that the pH of the solution did not change during the dissolution period of metal sulfides and CaS nanoparticles indicating that protons were neither consumed nor released during this time. In principle, this suggests a reaction of the form:

$$MS(s) + HS^-(aq) = MS(HS)^-(aq) \tag{8.3.1}$$

The dose of 2.0 g L^{-1} of CaS nanoparticles was found to be optimum for removal of heavy metals, where the removal efficiency reached 98–99.4%. It was also observed that Ni(II) removal efficiency was 96%. After treatment with CaS nanoparticles, the effluent heavy metals concentrations were below the industrial discharge limit. It was also observed that generated metal sulfide precipitate were colloidal in nature and took a significant time period to settle. The rate constant of the sulfide precipitation reaction for Fe(II), Co(II), Ni(II), Pb(II), Cd(II), Cu(II), and Zn(II) were 0.797, 0.831, 0.647, 0.090, 0.083, 0.00001, and 0.0828 min^{-1}, respectively. The lower K_f values may be due to the lower concentrations of (Cu(II), Pb(II), and Cd(II)) in the influent. Images of the continuous treatment of wastewater using the reactor(f) are shown in Figure B.2 in **Appendix** B.

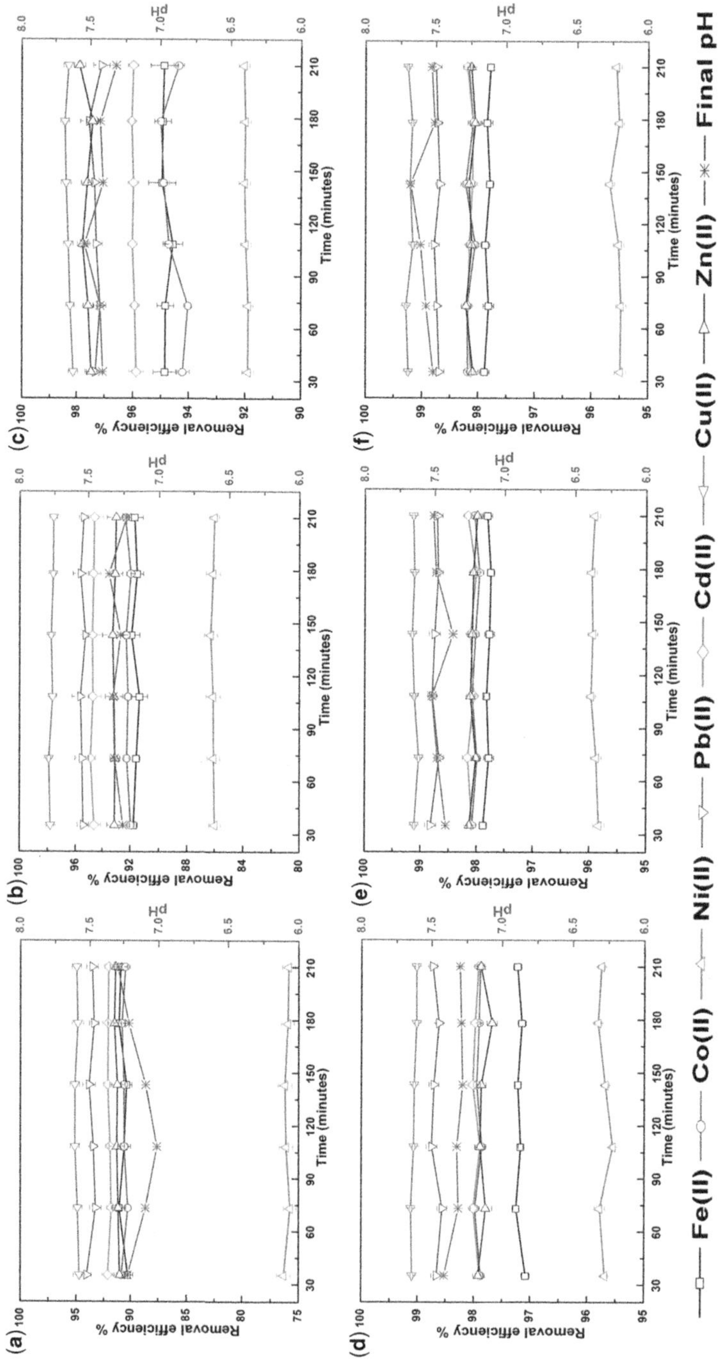

Figure 8.3 Dose dependent continuous study using CaS nanoparticles, 0.1 (a), 0.5 (b), 1.0 (c), 1.5 (d), 2.0 (e), and 2.5 (f) g L^{-1}.

8.3.3 Performance of stirred tank reactor (i) and Magnetic filtration unit (j)

Before performing continuous treatment using γ-Fe_2O_3 nanotubes, batch tests were performed to assess the specificity or selectivity of heavy metals adsorbed onto the adsorbent (γ-Fe_2O_3 nanotubes). From the batch test, it was found that the selectivity order of metal ions on to γ-Fe_2O_3 nanotubes was Cu(II) >Cd(II) >Zn(II) >Ni(II) >Co(II) >Pb(II) (Figure 8.4).

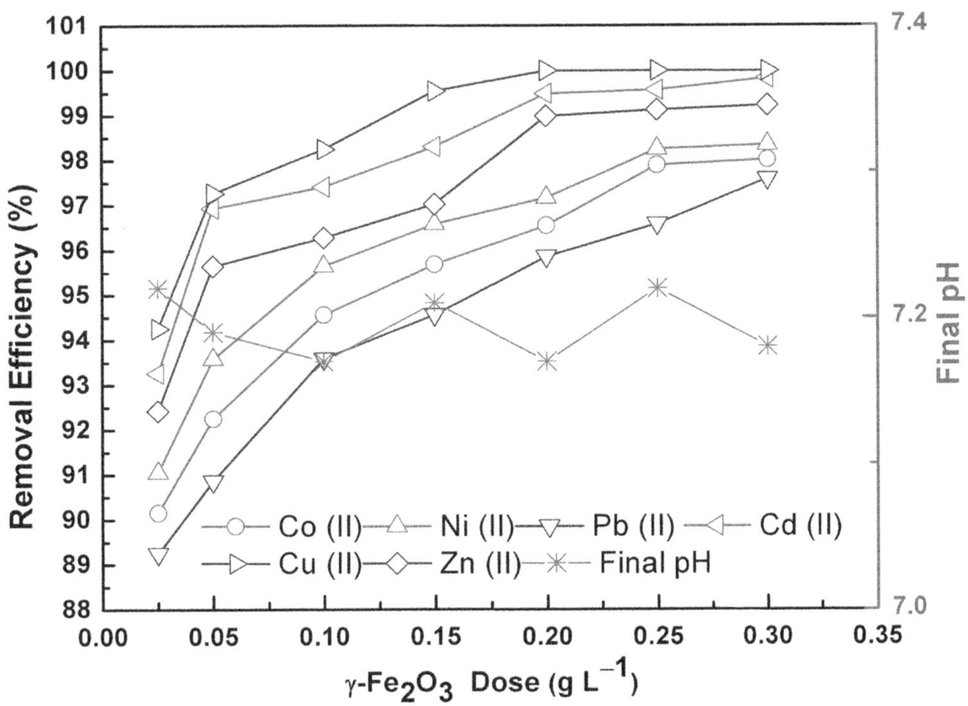

Figure 8.4 Heavy metal removals removal efficiency of γ-Fe_2O_3 nanotubes from mixed metals containing wastewater.

The effluent coming from reactor (f) treated with an optimum dose of 2.0 g L^{-1} of CaS nanoparticles, was used as influent for reactor (i) and magnetic filtration unit (j). The dose dependent heavy metal removal study was conducted using γ-Fe_2O_3 nanotubes, keeping it's flow rate and HRT constant. Figure 8.5 shows the performance of heavy metal removal using different dosages of γ-Fe_2O_3 nanotubes. The earlier study (Chapter 7), showed that the adsorption of heavy metal followed a Langmuir isotherm behavior and pseudo-second order kinetics. Magnetic nanotubes here act as adsorbent as well as a seed for settling of colloidal precipitates generated from the hydroxide and sulfide precipitation reactions (Figure 8.6).

It was also observed from Figure 8.5, the Fe(II) removal efficiency of magnetic nanotubes was lower than other metal removal efficiencies.

At a dose of 0.25 g L^{-1} of γ-Fe_2O_3 nanotubes it was found that Cu(II) was below the detection limit, and at dose of 0.3 g L^{-1}, not only Cu(II) but also Co(II) and Zn(II) were also below detection limit. From Figure 8.6 it was observed that using a dose of 0.3 g L^{-1} γ-Fe_2O_3 nanotubes, most of the metals removals approached 100% and some of them were below detection limits. Images of the continuous treatment of wastewater using the reactor(j) are given in Figure B.3 in **Appendix** B.

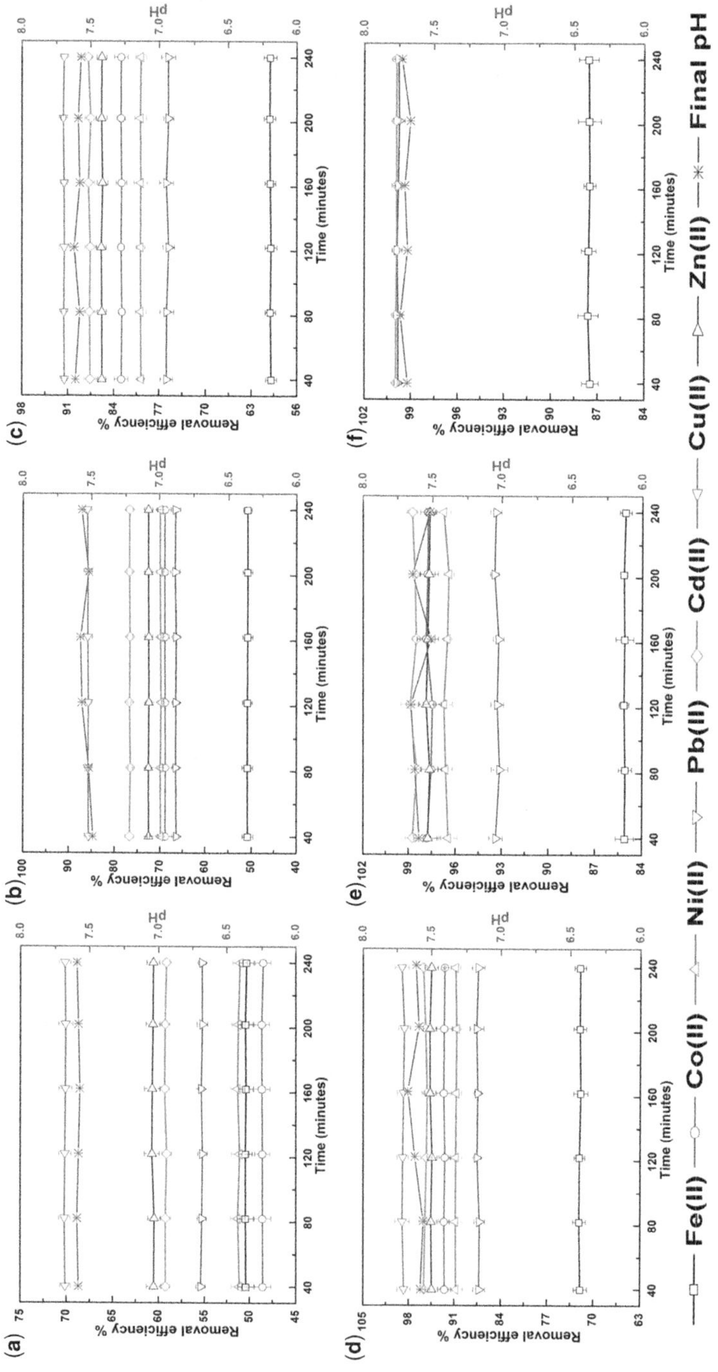

Figure 8.5 Dose dependent continuous study using γ-Fe_2O_3 nanotubes, 0.05 (a), 0.10 (b), 0.15 (c), 0.20 (d), 0.25 (e), and 0.30 (f) $g\ L^{-1}$.

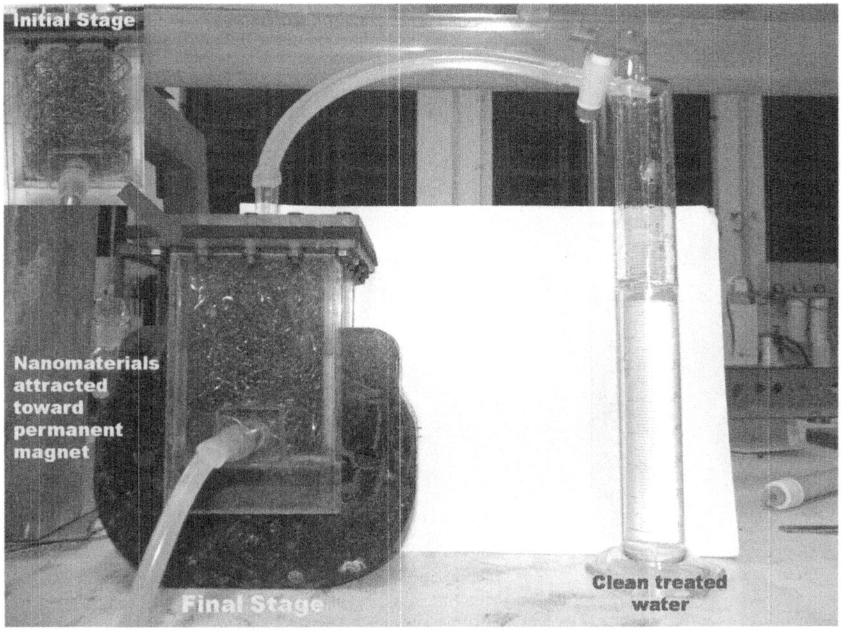

Figure 8.6 Magnetic filtration unit (j) arresting used magnetic adsorbent and colloidal precipitates.

8.3.4 Performance of multistage bench scale continuous reactor

On the basis of the semi-continuous treatment results, synthetic wastewater containing the same heavy metal load was treated continuously using three optimum doses of the different nanoparticles [1.5 g L^{-1} of $Ca(OH)_2$ dose at reactor (c), 2.0 g L^{-1} of CaS dose at reactor (f), 0.3 g L^{-1} of γ-Fe_2O_3 nanotubes dose at reactor (i)] with the predetermined flow rate and HRT. Table 8.3 lists the performance of the multistage bench scale continuous reactor and the final effluent was compared with running tap water in a laboratory.

From Table 8.3, it was also observed that at each settling tank, all metals were reduced due to precipitation settling. The final treated water containing Cd(II), Fe(II), Ni(II), and Pb(II) were below the USEPA (USEPA, 2011) drinking water level and Co(II), Cu(II), and Zn(II) were not detected. As the heavy metals were removed and precipitates were arrested using magnetic filtration unit, TDS, conductivity, and salinity also decreased and the values were comparable to tap water supply at IIT Kharagpur.

8.3.5 Economic feasibility study of multistage bench scale reactor

Conventional economic assessment of the experimented treatment scheme was performed. Table 8.4 lists the operating cost of the multistage bench scale continuous treatment method.

With a flow rate of 32 mL $min.^{-1}$ or 46 L day^{-1}, the working cost to treat 1.0 L with heavy metal load of 3750 mg L^{-1} wastewater would be Rs. 17/- or USD 0.28. The concentration of targeted heavy metals in treated water was not only below the limit of industrial discharge by EPA (CFR, 40), but was also below the limit for drinking water by USEPA (USEPA, 2011). These costs are calculated for one time use of the absorbing medium. Further, these operational costs can still be reduced if these treatment schemes are scaled to regular plant scale use.

Table 8.3 Multistage bench scale reactor results at specific condition.

Parameter	Reservoir (a)	Reactor (c)	Settling tank (d)	Reactor (f)	Settling tank (g)	Magnetic filter unit (j)	Tap Water
pH	1.99–2.12	7.60–7.80	7.50–7.60	7.40–7.50	7.40–7.50	7.45–7.55	7.35–7.50
TDS, (ppm)	2090 ± 0.05	1090 ± 0.03	1100 ± 0.02	877 ± 1.20	510 ± 1.50	226 ± 1.00	219 ± 1.50
Conductivity, (μS)	2940 ± 0.12	1540 ± 3.13	1452 ± 2.21	1036 ± 4.12	653 ± 2.12	323 ± 1.26	309 ± 2.13
Salinity, (ppm)	1700 ± 0.25	752 ± 2.31	685 ± 2.35	597 ± 2.14	441 ± 1.53	155 ± 1.32	146 ± 2.31
Heavy Metals, (ppm)							
Fe(II)	750	108 ± 2.0	105 ± 1.5	4.85 ± 2.5	4.01 ± 1.2	0.6 ± 0.012	1.2 ± 0.02
Cd(II)	500	155 ± 1.0	152 ± 2.0	5.32 ± 1.4	5.30 ± 0.2	0.001 ± 0.0001	ND
Co(II)	500	100 ± 2.2	99 ± 3.0	5.29 ± 2.1	5.26 ± 1.0	ND	ND
Cu(II)	500	55 ± 3.0	56 ± 1.0	1.03 ± 2.0	1.00 ± 0.5	ND	ND
Ni(II)	500	130 ± 1.5	127 ± 2.5	9.38 ± 3.1	9.35 ± 0.2	0.001 ± 0.0001	ND
Pb(II)	500	175 ± 2.0	174 ± 3.1	4.98 ± 2.6	4.86 ± 1.0	0.002 ± 0.0001	ND
Zn(II)	500	80 ± 1.4	81 ± 3.0	3.38 ± 2.0	3.36 ± 0.5	ND	ND
Ca(II)	10	420 ± 1.6	310 ± 1.2	330 ± 2.1	285 ± 0.5	280 ± 2.3	245 ± 2.01

N.B. : ± indicates standard error of the mean (n = 5) and ND = Not detected in AAS.

Table 8.4 Cost estimation of multistage bench scale treatment process.

		Operational cost		
Flow rate	**32 mL min.$^{-1}$ or 46 L day^{-1}**		**Influent heavy metals concentration**	**3,750 mg L^{-1}**
Consumable				
Influent heavy metals concentration		3,750 mg L^{-1}		
Influent heavy metals loading		120 mg min.$^{-1}$		
Consumable	**Consumption**	**Unit**	**Unit cost (Rs.)**	**Cost (Rs.) min.$^{-1}$**
$Ca(OH)_2$ nanoparticles	39	mg min.$^{-1}$	5.0 g^{-1}	0.195
CaS nanoparticle	32	mg min.$^{-1}$	7.5 g^{-1}	0.24
γ-Fe_2O_3	4.8	mg min.$^{-1}$	4.5 g^{-1}	0.0216
Total consumable				
Electric Power				
Power consumption	0.8	kWmin.	6.09 kWh^{-1}	0.08
Total operating cost (Rs.)				0.5366
Total cost to treat 1 L of wastewater (Rs.)				16.77

Appendices

Appendix A
Dose dependent study for commercial lime

A.1 DOSE DEPENDENT STUDY

The removal of heavy metals from synthetic effluent was investigated using commercial lime at nine dosage levels (0.1, 0.25, 0.5, 0.75, 1.0, 1.25, 1.5, 1.75 and 2.0 ($\pm$0.01) g L^{-1}). Tables A.1 revealed that increasing the dosage of commercial lime caused a rise in pH in the tested effluents. Dose dependent heavy metals removal efficiency from our experiments was shown in Figure A.1; these figure showed that the residual metals usually decreased as pH increased. Using lime dose of 2.0 g L^{-1}, pH of treated synthetic wastewater became 9.02–9.12. If dose increase there was no significant change observed in term of heavy metal removal.

At optimum dose (2.0 g L^{-1}) of commercial lime, heavy metal removal efficiency for different metals were found 80, 60, 80, 88, 67, 89, and 60% for Fe(II), Co(II), Ni(II), Pb(II), Cd(II), Cu(II), and Zn(II) respectively. In comparison with Ca-based nanoparticles, commercial lime showed lower heavy metals removal efficiency, due to may be lower dissolution of bulk nature of commercial lime. It was un able to react with metal ions present in synthetic wastewater.

Table A.1 Amount of heavy metal removal and final pH of the mixture after treatment with different dosages of lime.

Dosage ($g\ L^{-1}$)	Final pH	Fe(II) ($mg\ L^{-1}$) C_0[a]	C_f[b]	Cu(II) ($mg\ L^{-1}$) C_0[c]	C_f[d]	Pb(II) ($mg\ L^{-1}$) C_0[e]	C_f[f]	Cd(II) ($mg\ L^{-1}$) C_0[g]	C_f[h]	Ni(II) ($mg\ L^{-1}$) C_0[i]	C_f[j]	Co(II) ($mg\ L^{-1}$) C_0[k]	C_f[l]	Zn(II) ($mg\ L^{-1}$) C_0[m]	C_f[n]
0.10	3.12	99.12	87.17	99.38	88.15	97.80	89.25	97.73	91.53	99.84	89.22	99.82	91.01	99.40	91.24
0.25	3.99	99.14	79.95	99.34	72.27	97.78	81.28	97.70	83.45	99.80	78.55	99.81	87.38	99.40	86.08
0.50	4.69	99.12	73.76	99.41	68.21	97.81	76.34	97.69	78.38	99.89	69.85	99.82	81.01	99.41	80.55
0.75	5.59	99.13	62.91	99.35	55.15	97.82	61.83	97.75	71.06	99.87	61.02	99.80	76.22	99.38	76.54
1.00	5.97	99.16	58.85	99.35	40.03	97.80	53.22	97.74	65.54	99.81	53.15	99.85	68.64	99.39	64.01
1.25	7.19	99.14	47.25	99.32	35.33	97.93	44.01	97.70	57.06	99.85	45.89	99.83	60.02	99.43	58.21
1.50	7.98	99.15	34.46	99.34	23.27	97.91	34.08	97.71	41.27	99.85	36.41	99.87	54.85	99.40	50.09
1.75	8.66	99.11	22.75	99.36	11.73	97.87	21.18	97.76	37.69	99.82	27.23	99.85	49.97	99.45	46.82
2.00	9.02	99.15	19.61	99.40	10.51	97.88	11.81	97.73	31.82	99.83	20.08	99.84	39.82	99.44	40.55

[a] C_0 is initial concentration of heavy metal
[b] C_f is final concentration of heavy metal
[c] C_0 is initial concentration of heavy metal
[d] C_f is final concentration of heavy metal
[e] C_0 is initial concentration of heavy metal
[f] C_f is final concentration of heavy metal
[g] C_0 is initial concentration of heavy metal
[h] C_f is final concentration of heavy metal
[i] C_0 is initial concentration of heavy metal
[j] C_f is final concentration of heavy metal
[k] C_0 is initial concentration of heavy metal
[l] C_f is final concentration of heavy metal
[m] C_0 is initial concentration of heavy metal
[n] C_f is final concentration of heavy metal

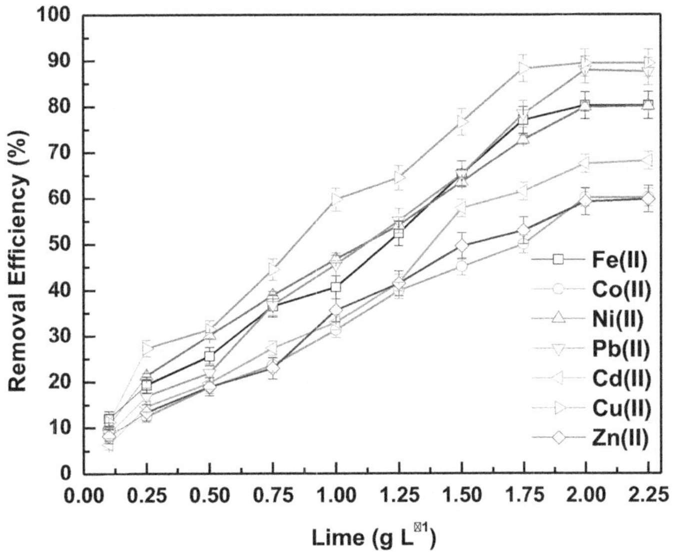

Figure A.1 Dose dependent heavy metals removal by Lime.

Appendix B

Continuous multistage bench–scale treatment of heavy metals from wastewater

Appendix B is also available in colour at http://www.iwaponline.com/wio/2014/Nanotechnology.pdf

B.1 IMAGE OF REACTOR (C)

Figure B.1 Time dependent continuous study of the reactor (c).

B.2 IMAGE OF REACTOR (F)

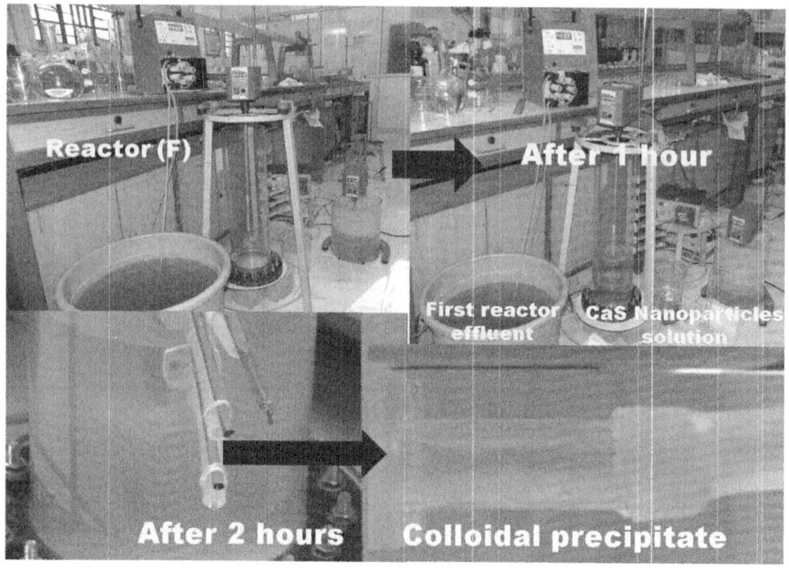

Figure B.2 Time dependent continuous study of the reactor (f).

B.3 IMAGE OF REACTOR (I)

Figure B.3 Time dependent continuous study of the reactor (i & j).

References

Adam D. (2003). Microwave chemistry: out of the kitchen. *Nature*, **421**(6923), 571–572.

Ahmad A. and Ooi B. (2010). A study on acid reclamation and copper recovery using low pressure nanofiltration membrane. *Chemical Engineering Journal*, **156**(2), 257–263.

Ahmad M. and Afzal H. (2001). Concentration levels of heavy and trace metals in the fish and relevant water from Rawal and Mangla lakes. *Journal of Biological Sciences*, **1**, 414–416.

Ai L. and Jiang J. (2009). Rapid synthesis of nanocrystalline co_3o_4 by a microwave-assisted combustion method. *Powder Technology*, **195**(1), 11–14.

Akbal F. and Camcı, S. (2010). Comparison of electrocoagulation and chemical coagulation for heavy metal removal. *Chemical Engineering & Technology*, **33**(10), 1655–1664.

Al-Asheh S. and Duvnjak Z. (1997). Sorption of cadmium and other heavy metals by pine bark. *Journal of Hazardous Materials*, **56**(1), 35–51.

Al-Jlil S. and Alsewailem F. (2009). Saudi arabian clays for lead removal in wastewater. *Applied Clay Science*, **42**(3), 671–674.

Al-llil S. (2010). Removal of heavy metals from industrial wastewater by adsorption using local Bentonite Clay and Roasted Date Pits in Saudi Arabia. *Trends in Applied Sciences Research*, **5**(2), 138–145.

Alloway B. (1995). *Heavy Metals in Soils*, Blackie Academic and Professional, London, Chapter Soil processes and the behavior of metals, pp. 38–57.

Ambrosi M., Dei L., Giorgi R., Neto C. and Baglioni P. (2001). Colloidal particles of $Ca(OH)_2$: properties and applications to restoration of frescoes. *Langmuir*, **17**(14), 4251–4255.

Amendola V. and Meneghetti M. (2009). Laser ablation synthesis in solution and size manipulation of noble metal nanoparticles. *Physical Chemistry Chemical Physics*, **11**(20), 3805–3821.

Anderson F., Dennis W. and Imbusch G. (2000). Transient luminescence line shapes for Mn^{2+} ions in ZnS nanocrystals. *Journal of Luminescence*, **90**(1), 27–32.

Annau Z. and Cuomo V. (1988). Mechanisms of neurotoxicity and their relationship to behavioral changes. *Toxicology*, **49**(2), 219–225.

Antonakos A., Liarokapis E. and Leventouri T. (2007). Micro-Raman and FTIR studies of synthetic and natural apatites. *Biomaterials*, **28**(19), 3043–3054.

APHA (1988). Standard Methods for the Examination of Water and Wastewater, 20th ed., Technical report, American Water Works Association and the Water Environment Federation, Washington, DC.

Apiratikul R. and Pavasant P. (2008). Sorption of Cu^{2+}, Cd^{2+}, and Pb^{2+} using modified zeolite from coal fly ash. *Chemical Engineering Journal*, **144**(2), 245–258.

Apostoli P. (1999). The role of element speciation in environmental and occupational medicine. *Fresenius' Journal of Analytical Chemistry*, **363**(5-6), 499–504.

Appell D. (2002). Nanotechnology: wired for success. *Nature*, **419**(6907), 553–555.

Artale M., Augugliaro V., Drioli E., Golemme G., Grande C., Loddo V., Molinari R., Palmisano L. and Schiavello M. (2000). Preparation and characterisation of membranes with entrapped TiO_2 and preliminary photocatalytic tests. *Annali di chimica*, **91**(3–4), 127–136.

Atieh M. (2011). Removal of chromium (VI) from polluted water using carbon nanotubes supported with activated carbon. *Procedia Environmental Sciences*, **4**, 281–293.

Aymonier C., Schlotterbeck U., Antonietti L., Zacharias P., Thomann R., Tiller J. and Mecking S. (2002). Hybrids of silver nanoparticles with amphiphilic hyperbranched macromolecules exhibiting antimicrobial properties. *Chemical Communications*, No. 24, 3018–3019.

Aziz H. A., Adlan M. N. and Ariffin K. S. (2008). Heavy metals (Cd, Pb, Zn, Ni, Cu and Cr(III)) removal from water in Malaysia: post treatment by high quality limestone. *Bioresource Technology*, **99**(6), 1578–1583.

Baghurst D. and Mingos D. (1992). Superheating effects associated with microwave dielectric heating. *Journal of the Chemical Society. Chemical communications*, No. 9, 674–677.

Bagwasi S., Tian B., Zhang J. and Nasir M. (2013). Synthesis, characterization and application of bismuth and boron co-doped TiO_2: a visible light active photocatalyst. *Chemical Engineering Journal*, **217**, 108–118.

Bagwe R. and Khilar K. (1997). Effects of the intermicellar exchange rate and cations on the size of silver chloride nanoparticles formed in reverse micelles of AOT. *Langmuir*, **13**(24), 6432–6438.

Baker C., Pradhan A., Pakstis L., Pochan D. and Shah S. (2005). Synthesis and antibacterial properties of silver nanoparticles. *Journal of Nanoscience and Nanotechnology*, **5**(2), 244–249.

Baltpurvins K., Burns R., Lawrance G. and Stuart A. (1997). Effect of electrolyte composition on zinc hydroxide precipitation by lime. *Water Research*, **31**(5), 973–980.

Barick K., Singh S., Aslam M. and Bahadur D. (2010). Porosity and photocatalytic studies of transition metal doped ZnO nanoclusters. *Microporous and Mesoporous Materials*, **134**(1), 195–202.

Barka-Bouaifel F., Sieber B., Bezzi N., Benner J., Roussel P., Boussekey L., Szunerits S. and Boukherroub R. (2011). Synthesis and photocatalytic activity of iodine-doped ZnO nanoflowers. *Journal of Materials Chemistry*, **21**(29), 10982–10989.

Barker T., Fatehi L., Lesnick M., Mealey T. and Raimond R. (2009). Nanotechnology and the Poor: Opportunities. *Nanotechnology & society*, Springer, pp. 243–263.

Barnes K., Liang J., Wu R., Worley S., Lee J., Broughton R. and Huang T. (2006). Synthesis and antimicrobial applications of 5, 5′-ethylenebis [5-methyl-3-(3-triethoxysilylpropyl) hydantoin], *Biomaterials*, **27**(27), 4825–4830.

Baruwati B., Kumar D. and Manorama S. (2006). Hydrothermal synthesis of highly crystalline ZnO nanoparticles: a competitive sensor for LPG and EtOH. *Sensors and Actuators B: Chemical*, **119**(2), 676–682.

Beliles R. (1979). The lesser metals. *Toxicity of Heavy Metals in the Environment, Part*, **2**, 383.

Belkacem M., Khodir M. and Abdelkrim S. (2008). Treatment characteristics of textile wastewater and removal of heavy metals using the electroflotation technique. *Desalination*, **228**(1), 245–254.

Bellama J., Jewett K., Manders W. and Nies J. (1988). A comparison of the rates of methylation of mercury (II) species in aquatic media by various organotin and organosilicon moieties. *Science of the Total Environment*, **73**(1), 39–51.

Bellobono I., Castellano L. and Tozzi A. (1991). Sulphur dioxide control by reactive photografted membranes immobilizing high surface area calcium oxide. *Materials Chemistry and Physics*, **28**(1), 69–74.

Bellobono I., Morazzoni F. and Tozzi P. (2005a). Photocatalytic membrane modules for drinking water purification in domestic and community appliances. *International Journal of Photoenergy*. **7**(3), 109–113.

Bellobono I., Morazzoni F., Bianchi R., Mangone E., Stanescu R., Costache C. and Tozzi P. (2005b). Solar energy driven photocatalytic membrane modules for water reuse in agricultural and food industries. pre-industrial experience using s-triazines as model molecules. *International Journal of Photoenergy*, **7**(2), 87–94.

Bellobono I., Selli E., Righetto L. and Muffato F. (1988). Flow dynamical characterization of sorbents immobilized as composites in membranes prepared by photochemical grafting onto polymers. *Materials Chemistry and Physics*, **19**(1–2), 131–146.

Betancur M., Bonelli P., Velásquez J. and Cukierman A. (2009). Potentiality of lignin from the Kraft pulping process for removal of trace nickel from wastewater: effect of demineralisation. *Bioresource Technology*, **100**(3), 1130–1137.

Bethune D. S., Kiang C. H., De Vries M. S., Gorman G., Savoy R., Vazquez J. and Beyers R. (1993). Cobalt-catalysed growth of carbon nanotubes with single-atomic-layer walls. *Nature*, **363**(6430), 605–607.

Bhat M., Chakravarthy B., Ramakrishnan P., Levasseur A. and Rao K. (2000). Microwave synthesis of electrode materials for lithium batteries, *Bulletin of Materials Science*, **23**(6), 461–466.

Bhatkhande D., Pangarkar V. and Beenackers A. (2002). Photocatalytic degradation for environmental applications-a review, *Journal of Chemical Technology and Biotechnology*, **77**(1), 102–116.

Bhattacharyya D., Jumawan Jr. A. and Grieves R. (1979). Separation of toxic heavy metals by sulfide precipitation. *Separation Science and Technology*, **14**(5), 441–452.

Biesaga M. and Pyrzynska K. (2006). The evaluation of carbon nanotubes as a sorbent for dicamba herbicide. *Journal of Separation Science*, **29**(14), 2241–2244.

Bilecka I., Kubli M., Amstad E. and Niederberger M. (2011). Simultaneous formation of ferrite nanocrystals and deposition of thin films via a microwave-assisted nonaqueous sol-gel process. *Journal of Sol-Gel Science and Technology*, **57**(3), 313–322.

BIS (1999). Tolerance Limits for Industrial Effluents Discharged into Inland Surface Waters. Technical Report IS 2490 (Parts 1), Bureau of Indian Standards, Manak Bhawan, New Delhi.

Blais J., Dufresne S. and Mercier G. (1999). État du développement technologique en matière d'enlèvement des métaux des effluents industriels. *Revue des sciences de l'eau/Journal of Water Science*, **12**(4), 687–711.

Bødker F., Hansen M., Koch C., Lefmann K. and Mørup S. (2000). Magnetic properties of hematite nanoparticles. *Physical Review B*, **61**(10), 6826.

Bogue R. (2008). Self-assembly: a review of recent developments. *Assembly Automation*, **28**(3), 211–215.

Bohnemann J., Libanori R., Moreira M. and Longo E. (2009). High-efficient microwave synthesis and characterisation of $SrSnO_3$. *Chemical Engineering Journal*, **155**(3), 905–909.

Bosc F., Ayral A. and Guizard C. (2005). Mesoporous anatase coatings for coupling membrane separation and photocatalyzed reactions. *Journal of Membrane Science*, **265**(1), 13–19.

Bowers C., Pietrass T., Barash E., Pines A., Grubbs R. and Alivisatos A. (1994). Probing CdS nanocrystal surfaces with laser-polarized xenon. *The Journal of Physical Chemistry*, **98**(38), 9400–9404.

Bratskaya S., Pestov A., Yatluk Y. and Avramenko V. (2009). Heavy metals removal by flocculation/precipitation using N-(2-carboxyethyl) chitosans. *Colloids and Surfaces A: Physicochemical and Engineering Aspects*, **339**(1), 140–144.

Brayner R., Ferrari-Iliou R., Brivois N., Djediat S., Benedetti M. and Fiévet F. (2006). Toxicological impact studies based on *Escherichia coli* bacteria in ultrafine ZnO nanoparticles colloidal medium. *Nano Letters*, **6**(4), 866–870.

Brewer M. and Scott T. (1983). Concise Encyclopedia Chemistry, Walter de Gruyter, Berlin, New York.

Brunauer S., Emmett P. and Teller E. (1938). Adsorption of gases in multimolecular layers. *Journal of the American Chemical Society*, **60**(2), 309–319.

Bykov Y., Eremeev A., Holoptsev V., Odemer C., Rachkovskii A. and Kleissl, H. (1997). Sintering of piezoceramics using millimeter-wave radiation. *Ceramic Transactions*, **80**: 321–328.

Bzdon S., Góralski J. and Maniukiewicz W., Perkowski J., Rogowski J. and Szadkowska-Nicze M. (2012). Radiation-induced synthesis of Fe-doped TiO_2: characterization and catalytic properties. *Radiation Physics and Chemistry*, **81**(3), 322–330.

Cai Y., Cai Y., Mou S. and Lu Y. (2005). Multi-walled carbon nanotubes as a solid-phase extraction adsorbent for the determination of chlorophenols in environmental water samples. *Journal of Chromatography A*, **1081**(2), 245–247.

Cai Y., Jiang G., Liu J. and Zhou Q. (2003a). Multiwalled carbon nanotubes as a solid-phase extraction adsorbent for the determination of bisphenol A, 4-n-nonylphenol, and 4-tert-octylphenol. *Analytical Chemistry*, **75**(10), 2517–2521.

Cai Y. Q., Jiang G. B., Liu J. F. and Zhou Q. X. (2003b). Multi-walled carbon nanotubes packed cartridge for the solid-phase extraction of several phthalate esters from water samples and their determination by high performance liquid chromatography. *Analytica Chimica Acta*, **494**(1), 149–156.

Cao C., Cui Z., Chen C., Song W. and Cai W. (2010). Ceria hollow nanospheres produced by a template-free microwave-assisted hydrothermal method for heavy metal ion removal and catalysis. *The Journal of Physical Chemistry C*, **114**(21), 9865–9870.

CCR (1985). Characteristic of Toxicity, Title 22, Division 4.5, Chapter 11, Article 3, Section 66261.24, Characteristic of Toxicity, Technical report, California Code of Regulations.

Cesar I., Kay A., Martinez J. and Grätzel M. (2006). Translucent thin film Fe_2O_3 photoanodes for efficient water splitting by sunlight: nanostructure-directing effect of Si-doping. *Journal of the American Chemical Society*, **128**(14), 4582–4583.

CFR (2003). Title 40, Chapter 1, Part 261.24, Toxicity Characteristic, Technical Report Part 261.24, US Code of Federal Regulations.

CFR (40). Code of Federal Regulations (CFR), 40, 141, 261, 268.40, Technical report, US Government Printing Office, Superintendent of Documents, Washington, DC.

Chamundeeswari M., Sobhana S., Jacob J., Kumar M., Devi M., Sastry T. and Mandal A. (2010). Preparation, characterization and evaluation of a biopolymeric gold nanocomposite with antimicrobial activity. *Biotechnology and Applied Biochemistry*, **55**(1), 29–35.

Chan T. and Reader H. (2000). Understanding Microwave Heating Cavities, Artech House Publishers.

Chang L. (1977). Neurotoxic effects of mercury–a review. *Environmental Research* **14**(3), 329–373.

Chang Q. and Wang G. (2007). Study on the macromolecular coagulant PEX which traps heavy metals. *Chemical Engineering Science*, **62**(17), 4636–4643.

Chang Q., Zhang M. and Wang J. (2009). Removal of Cu^{2+} and turbidity from wastewater by mercaptoacetyl chitosan. *Journal of Hazardous Materials*, **169**(1–3), 621–625.

Chang Y. and Chen D. (2005). Preparation and adsorption properties of monodisperse chitosan-bound Fe_3O_4 magnetic nanoparticles for removal of Cu(II) ions. *Journal of Colloid and Interface Science*, **283**(2), 446–451.

Charerntanyarak L. (1999). Heavy metals removal by chemical coagulation and precipitation. *Water Science and Technology*, **39**(10), 135–138.

Chawengkijwanich C. and Hayata Y. (2008). Development of TiO_2 powder-coated food packaging film and its ability to inactivate *Escherichia coli in vitro* and in actual tests. *International Journal of Food Microbiology*, **123**(3), 288–292.

Chen C. and Wang X. (2006). Adsorption of ni(II) from aqueous solution using oxidized multiwall carbon nanotubes. *Industrial & Engineering Chemistry Research*, **45**(26), 9144–9149.

Chen C., Hu J., Shao D., Li J. and Wang X. (2009a). Adsorption behavior of multiwall carbon nanotube/ iron oxide magnetic composites for Ni(II) and Sr(II). *Journal of Hazardous Materials*, **164**(2), 923–928.

Chen G. (2004). Electrochemical technologies in wastewater treatment. *Separation and Purification Technology*, **38**(1), 11–41.

Chen M., Cai Y., Yan Z. and Goodman D. (2006). On the origin of the unique properties of supported Au nanoparticles. *Journal of the American Chemical Society*, **128**(19), 6341–6346.

Chen Q., Luo Z., Hills C., Xue G. and Tyrer M. (2009b). Precipitation of heavy metals from wastewater using simulated flue gas: sequent additions of fly ash, lime and carbon dioxide. *Water Research*, **43**(10), 2605–2614.

Cheng D. (1989). Field and Wave Electromagnetics, Vol. 2, Addison-Wesley Tokyo.

Cheng X., Yu X., Xing Z. and Wan J. (2012). Enhanced photocatalytic activity of nitrogen doped TiO_2 anatase nano-particle under simulated sunlight irradiation. *Energy Procedia*, **16**, 598–605.

Chhabra V., Pillai V., Mishra B., Morrone A. and Shah D. (1995). Synthesis, characterization, and properties of microemulsion-mediated nanophase TiO_2 particles. *Langmuir*, **11**(9), 3307–3311.

Chiu H. and Yeh C. (2007). Hydrothermal synthesis of sno_2 nanoparticles and their gas-sensing of alcohol. *The Journal of Physical Chemistry C*, **111**(20), 7256–7259.

Choi H., Stathatos E. and Dionysiou D. (2006). So gel preparation of mesoporous photocatalytic TiO_2 films and TiO_2/Al_2O_3 composite membranes for environmental applications. *Applied Catalysis B-Environmental*, **63**(1), 60–67.

Choi H., Stathatos E. and Dionysiou D. (2007). Photocatalytic TiO_2 films and membranes for the development of efficient wastewater treatment and reuse systems. *Desalination*, **202**(1), 199–206.

Chou S., Krauss P., Zhang W., Guo L. and Zhuang L. (1997). Sub-10 nm imprint lithography and applications. *Journal of Vacuum Science & Technology B: Microelectronics and Nanometer Structures* **15**(6), 2897–2904.

Chu W. (1999). Lead metal removal by recycled alum sludge. *Water Research*, **33**(13), 3019–3025.

Cincinnati O. (1980). Control and Treatment Technology for the Metal Finishing Industry: Sulfide Precipitation, Summary report, The Industrial Environmental Research Laboratory.

Cioffi N., Torsi L., Ditaranto N., Tantillo G., Ghibelli L., Sabbatini L., Bleve-Zacheo T., D'Alessio M., Zambonin P. and Traversa E. (2005). Copper nanoparticle/polymer composites with antifungal and bacteriostatic properties. *Chemistry of Materials*, **17**(21), 5255–5262.

Clark D. and Sutton W. (1996). Microwave processing of materials. *Annual Review of Materials Science*, **26**(1), 299–331.

Clifford D. (1999). Ion Exchange and Inorganic Adsorption, Water Quality and Treatment, 5th edn, McGraw-Hill, New York.

Coey J. M. D. (2010). Magnetism and Magnetic Materials, Cambridge University Press, United States of America, New York.

Connor E., Mwamuka J., Gole A., Murphy C. and Wyatt M. (2005). Gold nanoparticles are taken up by human cells but do not cause acute cytotoxicity. *Small*, **1**(3), 325–327.

Couillard D. and Mercier G. (1992). Précipitations sélectives des métaux solubilisés biologiquement de boues aérobies d'épuration. *The Canadian Journal of Chemical Engineering*, **70**(5), 1021–1029.

Crane R. and Scott T. (2012). Nanoscale zero-valent iron: future prospects for an emerging water treatment technology. *Journal of Hazardous materials*, **211**, 112–125.

Cséfalvay E., Pauer V. and Mizsey P. (2009). Recovery of copper from process waters by nanofiltration and reverse osmosis. *Desalination*, **240**(1), 132–142.

Cullity B. and Stock S. (2001). Elements of X-ray Diffraction, Vol. 3, Prentice hall Upper Saddle River, NJ.

Dash S., Kamruddin M., Ajikumar P., Tyagi A. and Raj B. (2000). Nanocrystalline and metastable phase formation in vacuum thermal decomposition of calcium carbonate. *Thermochimica Acta*, **363**(1), 129–135.

Deliyanni E., Peleka E. and Matis K. (2007). Removal of zinc ion from water by sorption onto iron-based nanoadsorbent. *Journal of Hazardous Materials*, **141**(1), 176–184.

Diker H., Varlikli C., Mizrak K. and Dana A. (2011). Characterizations and photocatalytic activity comparisons of N-doped nc-TiO_2 depending on synthetic conditions and structural differences of amine sources. *Energy*, **36**(2), 1243–1254.

DIN DIN 38414-4, It is a standard. (1984). 38414 S4, Determination of leachability by water (S4), German standard methods for examination of water, wastewater and sludge, Sludge and Sediments (group S).

Dix H. (1981). Environmental Pollution, New York. John Wiley & Sons.

Dong F., Guo S., Wang H., Li X. and Wu Z. (2011). Enhancement of the visible light photocatalytic activity of C-doped TiO_2 nanomaterials prepared by a green synthetic approach. *The Journal of Physical Chemistry C*, **115**(27), 13285–13292.

Dowsett M. (1991). The application of surface analytical techniques to silicon technology. *Fresenius' Journal of Analytical Chemistry*, **341**(3–4), 224–234.

Duan J., Lu Q., Chen R., Duan Y., Wang L., Gao L. and Pan S. (2010). Synthesis of a novel flocculant on the basis of crosslinked konjac glucomannan-graft-polyacrylamide-co-sodium xanthate and its application in removal of Cu^{2+} ion. *Carbohydrate Polymers*, **80**(2), 436–441.

Duffus J. (2002). 'Heavy metals' a meaningless term? (IUPAC Technical Report). *Pure and Applied Chemistry*, **74**(5), 793–807.

Dyer J., Scrivner N. and Dentel S. (1998). A practical guide for determining the solubility of metal hydroxides and oxides in water. *Environmental Progress*, **17**(1), 1–8.

El-Awady M. and Sami T. (1997). Removal of heavy metals by cement kiln dust. *Bulletin of Environmental Contamination and Toxicology*, **59**(4), 603–610.

El Samrani A., Lartiges B. and Villiéras F. (2008). Chemical coagulation of combined sewer overflow: Heavy metal removal and treatment optimization. *Water Research*, **42**(4), 951–960.

ENV., E. C. D. (2002). E3 Project env. e.3/etu/0058, Heavy Metals in Waste. final report., *Technical report*, European Commission.

Estruga M., Domingo C. and Ayllón J. (2010). Microwave radiation as heating method in the synthesis of titanium dioxide nanoparticles from hexafluorotitanate-organic salts. *Materials Research Bulletin*, **45**(9), 1224–1229.

Fabrega J., Luoma S., Tyler C., Galloway T. and Lead J. (2011). Silver nanoparticles: behaviour and effects in the aquatic environment. *Environment International*, **37**(2), 517–531.

Fan H., Yi J., Yang Y., Kho K., Tan H., Shen Z., Ding J., Sun X., Olivo M. and Feng Y. (2009). Single-crystalline MFe_2O_4 nanotubes/nanorings synthesized by thermal transformation process for biological applications, *Acs Nano*, **3**(9), 2798–2808.

Faust B., Hoffmann M. and Bahnemann D. (1989). Photocatalytic oxidation of sulfur dioxide in aqueous suspensions of alpha-iron oxide (αFe_2O_3). *The Journal of Physical Chemistry*, **93**(17), 6371–6381.

Feng D., Aldrich C. and Tan H. (2000). Treatment of acid mine water by use of heavy metal precipitation and ion exchange. *Minerals Engineering*, **13**(6), 623–642.

Feng L., Cao M., Ma X., Zhu Y. and Hu C. (2012). Superparamagnetic high-surface-area Fe_3O_4 nanoparticles as adsorbents for arsenic removal. *Journal of Hazardous Materials*, **217**, 439–446.

Feynman R. (1959). There's Plenty of Room at the Bottom. *Engineering and Science* **23**(5), 22–36.

Figoli A., Cassano A., Criscuoli A., Mozumder M., Uddin M., Islam M. and Drioli E. (2010). Influence of operating parameters on the arsenic removal by nanofiltration. *Water Research*, **44**(1), 97–104.

for Toxic Substances A. and ATSDR D. R. (2000). Toxicological Profile for Arsenic, Technical report, Atlanta, GA: U.S. Department of Health and Human Services, Public Health Service.

Freundlich H. (1906). Über die adsorption in lösungen. *Z. Physik. Chemie (Leipzig)*, **57A**(385), 385–470.

Fu F. and Wang Q. (2011). Removal of heavy metal ions from wastewaters: a review. *Journal of Environmental Management*, **92**(3), 407–418.

Fugetsu B., Satoh S., Shiba T., Mizutani T., Lin Y., Terui N., Nodasaka Y., Sasa K., Shimizu K., Akasaka T., Shindoh M., Shibata K., Yokoyama A., Mori M., Tanaka K., Sato Y., Tohji K., Tanaka S., Nishi N. and Watari F. (2004). Caged multiwalled carbon nanotubes as the adsorbents for affinity-based elimination of ionic dyes. *Environmental Science & Technology*, **38**(24), 6890–6896.

Fujii T., de Groot F., Sawatzky G., Voogt F., Hibma T. and Okada K. (1999). In situ xps analysis of various iron oxide films grown by NO_2-assisted molecular-beam epitaxy. *Physical Review B*, **59**(4), 3195.

Galema S. (1997). Microwave chemistry. *Chemical Society reviews*, **26**(3), 233–238.

Gao C., Zhang W., Li H., Lang L. and Xu Z. (2008). Controllable fabrication of mesoporous MgO with various morphologies and their absorption performance for toxic pollutants in water, *Crystal Growth and Design*, **8**(10), 3785–3790.

Gaponov-Grekhov A. and Granatstein V. (1994). Applications of High-power Microwaves, Artech House Boston, London.

Gharabaghi M., Irannajad M. and Azadmehr A. (2012). Selective sulphide precipitation of heavy metals from acidic polymetallic aqueous solution by thioacetamide. *Industrial & Engineering Chemistry Research*, **51**(2), 954–963.

Giguere R., Namen A., Lopez B., Arepally A., Ramos D., Majetich G. and Defauw J. (1987). Studies on tandem ene/intramolecular diels-alder reactions. *Tetrahedron Letters*, **28**(52), 6553–6556.

Gode F. and Pehlivan E. (2006). Removal of chromium (III) from aqueous solutions using Lewatit S 100: the effect of pH, time, metal concentration and temperature. *Journal of Hazardous Materials*, **136**(2), 330–337.

Goldstein A., Gelb M. and Yager P. (2001). Continuous and highly variable rate controlled release of model drugs from sphingolipid-based complex high axial ratio microstructures. *Journal of Controlled Release*, **70**(1), 125–138.

González-Muñoz M., Rodríguez M., Luque S. and Álvarez J. (2006). Recovery of heavy metals from metal industry waste waters by chemical precipitation and nanofiltration. *Desalination*, **200**(1), 742–744.

Gonzalez P., Dominique Y., Massabuau J., Boudou A. and Bourdineaud J. (2005). Comparative effects of dietary methylmercury on gene expression in liver, skeletal muscle, and brain of the zebrafish (Danio rerio). *Environmental Science & Technology*, **39**(11), 3972–3980.

Gopalratnam V., Bennett G. and Peters R. (1992). Effect of collector dosage on metal removal by precipitation /flotation. *Journal of Environmental Engineering*, **118**(6), 923–948.

Goswami A., Raul P. and Purkait M. (2012). Arsenic adsorption using copper (II) oxide nanoparticles. *Chemical Engineering Research and Design*, **90**(9), 1387–1396.

Gotovac S., Hattori Y., Noguchi D., Miyamoto J., Kanamaru M., Utsumi S., Kanoh H. and Kaneko K. (2006). Phenanthrene adsorption from solution on single wall carbon nanotubes. *Journal of Physical Chemistry B*, **110**(33), 16219–16224.

Gracia-Pinilla M., Martínez E., Vidaurri G. and Pérez-Tijerina E. (2010). Deposition of size-selected Cu nanoparticles by inert gas condensation. *Nanoscale Research Letters*, **5**(1), 180–188.

Green A., Viney I., Brightwell J. and Ray B. (1988). Thermoluminescence emission spectra of single crystal CaS. *Materials Letters*, **7**(1–2), 19–21.

Gu X. and Evans L. (2008). Surface complexation modelling of Cd(II), Cu(II), Ni(II), Pb(II) and Zn(II) adsorption onto kaolinite. *Geochimica et Cosmochimica Acta* **72**(2), 267–276.

Gupta J., Barick K. and Bahadur D. (2011). Defect mediated photocatalytic activity in shape-controlled ZnOnanostructures. *Journal of Alloys and Compounds*, **509**(23), 6725–6730.

Gurunathan K., Maruthamuthu P. and Sastri M. (1997). Photocatalytic hydrogen production by dye-sensitized Pt/SnO_2 and Pt/SnO_2/RuO_2 in aqueous methyl viologen solution. *International Journal of Hydrogen Energy*, **22**(1), 57–62.

Haberzettl C. (2002). Nanomedicine: destination or journey? *Nanotechnology*, **13**, R9–R13.

Hakami O., Zhang Y. and Banks C. (2012). Thiol-functionalised mesoporous silica–coated magnetite nanoparticles for high efficiency removal and recovery of Hg from water. *Water Research*, **46**(12), 3913–3922.

Hall K., Eagleton L., Acrivos A. and Vermeulen T. (1966). Pore-and solid-diffusion kinetics in fixed-bed adsorption under constant-pattern conditions. *Industrial & Engineering Chemistry Fundamentals*, **5**(2), 212–223.

Hamada S. and Matijević, E. (1982). Formation of monodispersed colloidal cubic haematite particles in ethanol + water solutions. *Journal of the Chemical Society, Faraday Transactions 1: Physical Chemistry in Condensed Phases*, **78**(7), 2147–2156.

Hamada S., Kudo Y. and Minagawa K. (1990). The formation of monodispersed indium (III) hydroxide particles by forced hydrolysis at elevated temperature. *Bulletin of the Chemical Society of Japan*, **63**(1), 102–107.

Hammond C. and Hammond C. (2009). The Basics of Crystallography and Diffraction, 3rd edition, Oxford, United States, Oxford University Press Inc., New York.

Hankins N., Lu N. and Hilal N. (2006). Enhanced removal of heavy metal ions bound to humic acid by polyelectrolyte flocculation. *Separation and Purification Technology* **51**(1), 48–56.

Haswell S. (1991). Atomic Absorption Spectrometry: Theory, Design and Applications. Elsevier Science Publishers BV.

Heidmann I. and Calmano W. (2008). Removal of Zn(II), Cu(II), Ni(II), Ag(I) and Cr(VI) present in aqueous solutions by aluminium electrocoagulation. *Journal of Hazardous Materials*, **152**(3), 934–941.

Heredia J. and Martín J. (2009). Removing heavy metals from polluted surface water with a tannin-based flocculant agent. *Journal of Hazardous Materials*, **165**(1), 1215–1218.

Ho Y. and McKay G. (1999). Pseudo-second order model for sorption processes. *Process Biochemistry*, **34**(5), 451–465.

Hristovski K., Baumgardner A. and Westerhoff P. (2007). Selecting metal oxide nanomaterials for arsenic removal in fixed bed columns: from nanopowders to aggregated nanoparticle media. *Journal of Hazardous Materials*, **147**(1), 265–274.

Hsu W., Ronnquist L. and Matijevic E. (1988). Preparation and properties of monodispersed colloidal particles of lanthanide compounds. 2. Cerium(IV). *Langmuir*, **4**(1), 31–37.

Hu J., Chen G. and Lo I. (2005). Removal and recovery of Cr(VI) from wastewater by maghemite nanoparticles. *Water Research*, **39**(18), 4528–4536.

Hu J., Lo I. and Chen G. (2004). Removal of Cr(VI) by magnetite nanoparticle. *Water Science and Technology*, **50**(12), 139–146.

Huamain C., Chunrong Z., Cong T. and Yongguan Z. (1999). Heavy metal pollution in soils in China: status and countermeasures. *Ambio*, pp. 130–134.

Huang L., Li D., Lin Y., Wei M., Evans D. and Duan X. (2005). Controllable preparation of Nano-MgO and investigation of its bactericidal properties. *Journal of Inorganic Biochemistry*, **99**(5), 986–993.

Huisman J., Schouten G. and Schultz C. (2006). Biologically produced sulphide for purification of process streams, effluent treatment and recovery of metals in the metal and mining industry, *Hydrometallurgy*, **83**(1), 106–113.

Humphrey M., Cole M., Pendergrass J. and Kiningham K. (2005). Mitochondrial mediated thimerosal-induced apoptosis in a human neuroblastoma cell line (SK-N-SH). *Neurotoxicology*, **26**(3), 407–416.

Iannelli M., Alupei V. and Ritter H. (2005). Selective microwave-accelerated synthesis and polymerization of chiral methacrylamide directly from methacrylic acid and (*R*)-1-phenyl-ethylamine. *Tetrahedron*, **61**(6), 1509–1515.

Iijima S. (1991). Helical microtubules of graphitic carbon. *Nature*, **354**(6348), 56–58.

Ireland J., Klostermann P., Rice E. and Clark R. (1993). Inactivation of *Escherichia coli* by titanium dioxide photocatalytic oxidation. *Applied and Environmental Microbiology*, **59**(5), 1668–1670.

Jajarmi P. (2009). Fabrication of pure ZnO nanoparticles by polymerization method. *Materials Letters*, **63**(30), 2646–2648.

James S., Slikker III, W., Melnyk S., New E., Pogribna M. and Jernigan S. (2005). Thimerosal neurotoxicity is associated with glutathione depletion: protection with glutathione precursors. *Neurotoxicology*, **26**(1), 1–8.

Janney M. (1988). Microwave sintering of alumina at 28 GHz. *Ceramic Powder Science II. Transactions Westerville, Oh., 1988,*, **1**: 919–924.

Jenkins R. and Snyder R. (2012). Introduction to X-ray Powder Diffractometry, Vol. 267, Wiley-Interscience.

Jeon C., Park J. and Yoo Y. (2001). Removal of heavy metals in plating wastewater using carboxylated alginic acid. *Korean Journal of Chemical Engineering*, **18**(6), 955–960.

Jia C., Sun L., Yan Z., You L., Luo F., Han X., Pang Y., Zhang Z. and Yan C. (2005). Single-crystalline iron oxide nanotubes. *Angewandte Chemie*, **117**(28), 4402–4407.

Jiang W., Mashayekhi H. and Xing B. (2009). Bacterial toxicity comparison between nano-and micro-scaled oxide particles. *Environmental Pollution*, **157**(5), 1619–1625.

Jin M., Kim N., Kim H., Yoon C., Lee C., Kim M. and Kim W. (2001). Optical properties of undoped and Co^{2+}-Doped CaS, CaSe, BaS, and BaSe single crystals. *Journal Korean Physical Society*, **39**(1), 692–697.

Jin T., Sun D., Su J., Zhang H. and Sue H. (2009). Antimicrobial efficacy of zinc oxide quantum dots against *Listeria monocytogenes, Salmonella enteritidis, and Escherichia coli* O157: H7. *Journal of Food Science*, **74**(1), M46–M52.

Johnson S., Franks G., Scales P., Boger D. and Healy T. (2000). Surface chemistry-rheology relationships in concentrated mineral suspensions. *International Journal of Mineral Processing*, **58**(1), 267–304.

Jong T. and Parry D. (2003). Removal of sulfate and heavy metals by sulfate reducing bacteria in short-term bench scale upflow anaerobic packed bed reactor runs. *Water Research*, **37**(14), 3379–3389.

Jouhannaud J., Rossignol J. and Stuerga D. (2008). Rapid synthesis of tin (IV) oxide nanoparticles by microwave induced thermohydrolysis. *Journal of Solid State Chemistry*, **181**(6), 1439–1444.

Ju J., Chen X., Shi Y., Miao J. and Wu D. (2013). Hydrothermal preparation and photocatalytic performance of N, S-doped nanometer TiO_2 under sunshine irradiation. *Powder Technology*, **237**: 616–622.

Jusoh A., Shiung L., Ali N. and Noor M. (2007). A simulation study of the removal efficiency of granular activated carbon on cadmium and lead. *Desalination*, **206**(1), 9–16.

Jüttner K., Galla U. and Schmieder H. (2000). Electrochemical approaches to environmental problems in the process industry. *Electrochimica Acta*, **45**(15), 2575–2594.

Kabata-Pendias A. (2000). Trace Elements in Soils and Plants, CRC press.

Kabbashi N., Atieh M., Al-Mamun A., Mirghami M., Alam M. and Yahya N. (2009). Kinetic adsorption of application of carbon nanotubes for Pb(II) removal from aqueous solution. *Journal of Environmental Sciences*, **21**(4), 539–544.

Kadirvelu K., Thamaraiselvi K. and Namasivayam C. (2001). Removal of heavy metals from industrial wastewaters by adsorption onto activated carbon prepared from an agricultural solid waste. *Bioresource Technology*, **76**(1), 63–65.

Kandah M. and Meunier J. (2007). Removal of nickel ions from water by multi-walled carbon nanotubes, *Journal of Hazardous Materials*, **146**(1–2), 283–288.

Kang K., Kim S., Choi J. and Kwon S. (2008). Sorption of Cu^{2+} and Cd^{2+} onto acid-and base-pretreated granular activated carbon and activated carbon fiber samples. *Journal of Industrial and Engineering Chemistry*, **14**(1), 131–135.

Kappe C. (2002). High-speed combinatorial synthesis utilizing microwave irradiation, *Current Opinion in Chemical Biology*, **6**(3), 314–320.

Kappe C. (2004). Controlled microwave heating in modern organic synthesis. *Angewandte Chemie International Edition*, **43**(46), 6250–6284.

Kerner R., Palchik O. and Gedanken A. (2001). Sonochemical and microwave-assisted preparations of PbTe and PbSe. A comparative study. *Chemistry of Materials*, **13**(4), 1413–1419.

Khizroev S., Kryder M., Litvinov D. and Thompson D. (2002). Direct observation of magnetization switching in focused-ion-beam-fabricated magnetic nanotubes. *Applied Physics Letters*, **81**(12), 2256–2257.

Khosravi A., Kundu M., Jatwa L., Deshpande S., Bhagwat U., Sastry M. and Kulkarni S. (1995). Green luminescence from copper doped zinc sulphide quantum particles, *Applied Physics Letters*, **67**(18), 2702–2704.

Kim B., Kim D., Cho D. and Cho S. (2003a). Bactericidal effect of TiO_2 photocatalyst on selected foodborne pathogenic bacteria. *Chemosphere*, **52**(1), 277–281.

Kim D. and Huh Y. (2011). Morphology-dependent photocatalytic activities of hierarchical microstructures of ZnO. *Materials Letters*, **65**(14), 2100–2103.

Kim J., Kuk E., Yu K. N., Kim J., Park S., Lee H. J., Kim S., Park Y., Park Y., Hwang C. *et al.* (2007). Antimicrobial effects of silver nanoparticles. *Nanomedicine: Nanotechnology, Biology and Medicine*, **3**(1), 95–101.

Kim S., Kwak S., Sohn B. and Park T. (2003b). Design of TiO_2 nanoparticle self-assembled aromatic polyamide thin-film-composite (TFC) membrane as an approach to solve biofouling problem. *Journal of Membrane Science*, **211**(1), 157–165.

Klabunde K., Stark J., Koper O., Mohs C., Park D., Decker S., Jiang Y., Lagadic I. and Zhang D. (1996). Nanocrystals as stoichiometric reagents with unique surface chemistry. *The Journal of Physical Chemistry*, **100**(30), 12142–12153.

Kleine J., Peinemann K., Schuster C. and Warnecke H. (2002). Multifunctional system for treatment of wastewaters from adhesive-producing industries: separation of solids and oxidation of dissolved pollutants using doted microfiltration membranes. *Chemical Engineering Science*, **57**(9), 1661–1664.

Köhler S., Cubillas P., Rodríguez-Blanco J., Bauer C. and Prieto, M. (2007). Removal of cadmium from wastewaters by aragonite shells and the influence of other divalent cations. *Environmental Science & Technology*, **41**(1), 112–118.

Kongsricharoern N. and Polprasert C. (1995). Electrochemical precipitation of chromium (Cr^{6+}) from an electroplating wastewater. *Water Science and Technology*, **31**(9), 109–117.

Konishi M., Isobe T. and Senna M. (2001). Enhancement of photoluminescence of ZnS: Mn nanocrystals by hybridizing with polymerized acrylic acid. *Journal of luminescence*, **93**(1), 1–8.

Koper O., Klabunde J., Marchin G., Klabunde K., Stoimenov P. and Bohra L. (2002). Nanoscale powders and formulations with biocidal activity toward spores and vegetative cells of *bacillus* species, viruses, and toxins. *Current Microbiology*, **44**(1), 49–55.

Koper O., Lagadic I., Volodin A. and Klabunde K. (1997). Alkaline-earth oxide nanoparticles obtained by aerogel methods. Characterization and rational for unexpectedly high surface chemical reactivities. *Chemistry of Materials*, **9**(11), 2468–2480.

Koper O., Li Y. and Klabunde K. (1993). Destructive adsorption of chlorinated hydrocarbons on ultrafine (nanoscale) particles of calcium oxide. *Chemistry of Materials*, **5**(4), 500–505.

Ku Y. and Peters R. (1987). Innovative uses from carbon adsorption of heavy metals from plating wastewaters: I. Activated carbon polishing treatment. *Environmental Progress*, **6**(2), 119–124.

Kuo C. and Lin H. (2009). Adsorption of aqueous cadmium (II) onto modified multi-walled carbon nanotubes following microwave/chemical treatment. *Desalination*, **249**(2), 792–796.

Kurihara K., Kizling J., Stenius P. and Fendler J. (1983). Laser and pulse radiolytically induced colloidal gold formation in water and in water-in-oil microemulsions. *Journal of the American Chemical Society*, **105**(9), 2574–2579.

Kurniawan T., Chan G., Lo W. and Babel S. (2006). Physico-chemical treatment techniques for wastewater laden with heavy metals. *Chemical Engineering Journal*, **118**(1), 83–98.

Kwak S., Kim S. and Kim S. (2001). Hybrid organic/inorganic reverse osmosis (RO) membrane for bactericidal anti-fouling. 1. preparation and characterization of TiO_2 nanoparticle self-assembled aromatic polyamide thin-film-composite (TFC) membrane. *Environmental Science & Technology*, **35**(11), 2388–2394.

Lagergren S. (1898). Zur theorie der sogenannten adsorption geloster stoffe. kungliga svenska vetenskapsakademiens. *Handlingar*, **24**: 1–39.

Lan H., Ding Y., Liu H. and Lu B. (2007). Review of the wafer stage for nanoimprint lithography. *Microelectronic Engineering*, **84**(4), 684–688.

Langmuir I. (1916). The constitution and fundamental properties of solids and liquids. Part I. Solids. *Journal of the American Chemical society*, **38**(2), 2221–2295.

Lankford P. and Eckenfelder W. (1990). Toxicity Reduction in Industrial Effluents, New York, NY (United States); Van Nostrand Reinhold.

Lara H., Ayala-Núñez N., Turrent L. and Padilla C. (2010). Bactericidal effect of silver nanoparticles against multidrug-resistant bacteria. *World Journal of Microbiology and Biotechnology*, **26**(4), 615–621.

Lee M., Jung B., Chung S., Bae O., Lee J., Park J., Yang J., Lee H. and Chung J. (2003). Arsenic-induced dysfunction in relaxation of blood vessels. *Environmental Health Perspectives*, **111**(4), 513.

Lewis A. (2010). Review of metal sulphide precipitation. *Hydrometallurgy*, **104**(2), 222–234.

Li B. and Wang Y. (2009). Facile synthesis and enhanced photocatalytic performance of flower-like ZnO hierarchical microstructures. *The Journal of Physical Chemistry C*, **114**(2), 890–896.

Li L., Lu J., Wang Z., Yang L., Zhou X. and Han L. (2012). Fabrication of the CN co-doped rod-like TiO_2 photocatalyst with visible-light responsive photocatalytic activity. *Materials Research Bulletin*, **47**(6), 1508–1512.

Li X., Wang L. and Lu X. (2010a). Preparation of silver-modified TiO_2 via microwave-assisted method and its photocatalytic activity for toluene degradation. *Journal of Hazardous Materials*, **177**(1), 639–647.

Li Y., Gong J. and Deng Y. (2010b). Hierarchical structured ZnO nanorods on ZnO nanofibers and their photoresponse to UV and visible lights. *Sensors and Actuators A: Physical*, **158**(2), 176–182.

Li Y. H., Wang S., Luan Z., Ding J., Xu C. and Wu D. (2003). Adsorption of cadmium (II) from aqueous solution by surface oxidized carbon nanotubes. *Carbon*, **41**(5), 1057–1062.

Li Y., Liu F., Xia B., Du Q., Zhang P., Wang D., Wang Z. and Xia Y. (2010c). Removal of copper from aqueous solution by carbon nanotube\calcium alginate composites. *Journal of Hazardous Materials*, **177**(1), 876–880.

Li Y., Wang S., Wei J., Zhang X., Xu C., Luan Z., Wu D. and Wei B. (2002). Lead adsorption on carbon nanotubes. *Chemical Physics Letters*, **357**(3), 263–266.

Li Y., Zhu Y., Zhao Y., Wu D. and Luan Z. (2006). Different morphologies of carbon nanotubes effect on the lead removal from aqueous solution. *Diamond and Related Materials*, **15**(1), 90–94.

Liang P., Ding Q. and Song F. (2005). Application of multiwalled carbon nanotubes as solid phase extraction sorbent for preconcentration of trace copper in water samples. *Journal of Separation Science*, **28**(17), 2339–2343.

Liang P., Liu Y., Guo L., Zeng J. and Lu H. (2004). Multiwalled carbon nanotubes as solid-phase extraction adsorbent for the preconcentration of trace metal ions and their determination by inductively coupled plasma atomic emission spectrometry. *Journal of Analytical Atomic Spectrometry*, **19**(11), 1489–1492.

Liao X., Chen N., Xu S., Yang S. and Zhu J. (2003). A microwave assisted heating method for the preparation of copper sulfide nanorods. *Journal of Crystal Growth*, **252**(4), 593–598.

Lin S., Lai S. and Leu H. (2000). Removal of heavy metals from aqueous solution by chelating resin in a multistage adsorption process. *Journal of Hazardous Materials*, **76**(1), 139–153.

Link G., Ivanov V., Paranin S., Khrustov V., Böhme R., Müller G., Schumacher G., Thumm M. and Weisenburger A. (1996). A comparison of mm-wave sintering and fast conventional sintering of nanocrystalline Al_2O_3, MRS Proceedings, Vol. 430, Cambridge Univ Press.

Lisiecki I. and Pileni M. (1995). Copper metallic particles synthesized "in situ" in reverse micelles: influence of various parameters on the size of the particles. *The Journal of Physical Chemistry*, **99**(14), 5077–5082.

Liu G., Han C., Pelaez M., Zhu D., Liao S., Likodimos V., Ioannidis N., Kontos A., Falaras P., Dunlop P., Byrne J. and Dionysiou D. (2012). Synthesis, characterization and photocatalytic evaluation of visible light activated C–doped TiO_2 nanoparticles. *Nanotechnology*, **23**(29), 294003.

Liu J., Zhao Z. and Jiang G. (2008a). Coating Fe_3O_4 magnetic nanoparticles with humic acid for high efficient removal of heavy metals in water. *Environmental Science & Technology*, **42**(18), 6949–6954.

Liu Y., He L., Mustapha A., Li H., Hu Z. and Lin M. (2009). Antibacterial activities of zinc oxide nanoparticles against *Escherichia coli* O157: H7. *Journal of Applied Microbiology* **107**(4), 1193–1201.

Liu Z., Zhang D., Han S., Li C., Lei B., Lu W., Fang J. and Zhou C. (2005). Single crystalline magnetite nanotubes. *Journal of the American Chemical Society*, **127**(1), 6–7.

Liu Z., Zhou L., Wei P., Zeng K., Wen C. and Lan H. (2008b). Competitive adsorption of heavy metal ions on peat. *Journal of China University of Mining and Technology*, **18**(2), 255–260.

Long R. Q. and Yang R. T. (2001). Carbon nanotubes as superior sorbent for dioxin removal. *Journal of the american chemical society*, **123**(9), 2058–2059.

Lozet J. and Mathieu C. (1990). Dictionary of Soil Science, number Ed 2, Technique et Documentation-Lavoisier.

Lu C. and Liu C. (2006). Removal of nickel (II) from aqueous solution by carbon nanotubes, *Journal of Chemical Technology and Biotechnology*. **81**(12), 1932–1940.

Lu C., Chung Y. L. and Chang K. F. (2005). Adsorption of trihalomethanes from water with carbon nanotubes. *Water Research*, **39**(6), 1183–1189.

Lu C., Chung Y. L. and Chang K. F. (2006). Adsorption thermodynamic and kinetic studies of trihalomethanes on multiwalled carbon nanotubes. *Journal of Hazardous Materials*, **138**(2), 304–310.

Lundh M., Jönsson L. and Dahlquist J. (2000). Experimental studies of the fluid dynamics in the separation zone in dissolved air flotation. *Water Research*, **34**(1), 21–30.

Luo T., Cui J., Hu S., Huang Y. and Jing C. (2010). Arsenic removal and recovery from copper smelting wastewater using TiO_2. *Environmental Science & Technology*, **44**(23), 9094–9098.

Lv B., Xu Y., Wu D. and Sun Y. (2008). Preparation and properties of magnetic iron oxide nanotubes. *Particuology*, **6**(5), 334–339.

Lv C., Zhou Y., Li H., Dang M., Guo C., Ou Y. and Xiao B. (2011). Synthesis and characterisation of Gd^{3+}-doped mesoporous TiO_2 materials. *Applied Surface Science*, **257**(11), 5104–5108.

Ma S., Li R., Lv C., Xu W. and Gou X. (2011). Facile synthesis of ZnO nanorod arrays and hierarchical nanostructures for photocatalysis and gas sensor applications. *Journal of Hazardous Materials*, **192**(2), 730–740.

Ma Y., Vileno E., Suib S. and Dutta P. (1997). Synthesis of tetragonal $batio_3$ by microwave heating and conventional heating. *Chemistry of Materials*, **9**(12), 3023–3031.

Mafuné, F., Kohno J., Takeda Y. and Kondow T. (2002). Full physical preparation of size-selected gold nanoparticles in solution: laser ablation and laser-induced size control. *The Journal of Physical Chemistry B*, **106**(31), 7575–7577.

Mahapatra O., Bhagat M., Gopalakrishnan C. and Arunachalam K. (2008). Ultrafine dispersed cuo nanoparticles and their antibacterial activity. *Journal of Experimental Nanoscience*, **3**(3), 185–193.

Malkoc E., Nuhoglu Y. and Dundar M. (2006). Adsorption of chromium (VI) on pomace–an olive oil industry waste: batch and column studies. *Journal of Hazardous Materials*, **138**(1), 142–151.

Marchioretto M., Bruning H. and Rulkens W. (2005). Heavy metals precipitation in sewage sludge. *Separation Science and Technology*, **40**(16), 3393–3405.

Martinez-Gutierrez F., Olive P., Banuelos A., Orrantia E., Nino N., Sanchez E., Ruiz F., Bach H. and Av-Gay Y. (2010). Synthesis, characterization, and evaluation of antimicrobial and cytotoxic effect of silver and titanium nanoparticles. *Nanomedicine: Nanotechnology, Biology and Medicine*, **6**(5), 681–688.

Matijević, E. and Scheiner P. (1978). Ferric hydrous oxide sols: III. Preparation of uniform particles by hydrolysis of Fe (III)-chloride,-nitrate, and-perchlorate solutions. *Journal of Colloid and Interface Science*, **63**(3), 509–524.

Matis K., Zouboulis A., Gallios G., Erwe T. and Blöcher C. (2004). Application of flotation for the separation of metal-loaded zeolites. *Chemosphere*, **55**(1), 65–72.

Matlock M., Howerton B. and Atwood D. (2002). Chemical precipitation of heavy metals from acid mine drainage. *Water Research*, **36**(19), 4757–4764.

Matsunaga T., Tomoda R., Nakajima T. and Wake H. (1985). Photoelectrochemical sterilization of microbial cells by semiconductor powders. *FEMS Microbiology Letters*, **29**(1), 211–214.

Matsunaga T., Tomoda R., Nakajima T., Nakamura N. and Komine T. (1988). Continuous-sterilization system that uses photosemiconductor powders. *Applied and Environmental Microbiology*, **54**(6), 1330–1333.

Meek T., Blake R. and Petrovic J. (1987). Microwave sintering of Al_2O_3 and Al_2O_3-SiC whisker composites, 11th Annual Conference on Composites and Advanced Ceramic Materials: Ceramic Engineering and Science Proceedings, Wiley Online Library, pp. 861–871.

Metcalf L., Eddy H. and Tchobanoglous G. (2010). *Wastewater engineering: treatment, disposal, and reuse*, McGraw-Hill.

Meulenkamp E. (1998). Size dependence of the dissolution of ZnO nanoparticles. *The Journal of Physical Chemistry B*, **102**(40), 7764–7769.

Meunier N., Drogui P., Montané, C., Hausler R., Blais J. and Mercier, G. (2006). Heavy metals removal from acidic and saline soil leachate using either electrochemical coagulation or chemical precipitation. *Journal of Environmental Engineering*, **132**(5), 545–554.

Michel P. (1993). Arsenic in marine medium; biochemistry and ecotoxicology, pepères, Ocean, Ifremer, Paris.

Migdisov A., Williams-Jones A., Lakshtanov L. and Alekhin Y. (2002). Estimates of the second dissociation constant of H_2S from the surface sulfidation of crystalline sulfur. *Geochimica et Cosmochimica Acta*, **66**(10), 1713–1725.

Mingos D. (1994). The applications of microwaves in chemical syntheses. *Research on Chemical Intermediates*, **20**(1), 85–91.

Mirbagheri S. and Hosseini S. (2005). Pilot plant investigation on petrochemical wastewater treatmentfor the removal of copper and chromium with the objective of reuse. *Desalination*, **171**(1), 85–93.

Mirkin C. and Taton T. (2000). Materials chemistry: semiconductors meet biology. *Nature*, **405** (6787), 626–627.

Mishra V. and Kumar A. (2009). Impact of metal nanoparticles on the plant growth promoting rhizobacteria. *Digest Journal of Nanomaterials and Biostructures*, **4**: 587–592.

Mohammad G., Mishra V. and Pandey H. (2008). Antioxidant properties of some nanoparticle may enhance wound healing in T2DM patient. *Digest Journal of Nanomaterials and Biostructures* **3**: 159–162.

Mohan D. and Chander S. (2006). Removal and recovery of metal ions from acid mine drainage using lignite-a low cost sorbent. *Journal of Hazardous Materials*, **137**(3), 1545–1553.

Mohan R., Krishnamoorthy K. and Kim S. (2012). Enhanced photocatalytic activity of Cu-doped ZnO nanorods. *Solid State Communications*, **152**(5), 375–380.

Mohsen-Nia M., Montazeri P. and Modarress H. (2007). Removal of Cu^{2+} and Ni^{2+} from wastewater with a chelating agent and reverse osmosis processes. *Desalination*, **217**(1), 276–281.

Molinari R., Pirillo F., Falco M., Loddo V. and Palmisano L. (2004). Photocatalytic degradation of dyes by using a membrane reactor. *Chemical Engineering and Processing: Process Intensification*, **43**(9), 1103–1114.

Morken T. S., Sonnewald U., Aschner M. and Syversen T. (2005). Effects of methylmercury on primary brain cells in mono-and co-culture. *Toxicological Sciences*, **87**(1), 169–175.

Morris C. (1992). Academic Press Dictionary of Science and Technology [Electronic Book], Academic Press, San Diego.

Motsi T., Rowson N. and Simmons M. (2009). Adsorption of heavy metals from acid mine drainage by natural zeolite. *International Journal of Mineral Processing*, **92**(1), 42–48.

Muñoz J., Gallego M. and Valcárcel M. (2005). Speciation of organometallic compounds in environmetal samples by gas chromatography after flow preconcentration on fullerenes and nanotubes. *Analytical Chemistry*, **77**(16), 5389–5395.

Murthy Z. and Chaudhari L. (2008). Application of nanofiltration for the rejection of nickel ions from aqueous solutions and estimation of membrane transport parameters. *Journal of Hazardous Materials*, **160**(1), 70–77.

Muthukrishnan M. and Guha B. (2008). Effect of pH on rejection of hexavalent chromium by nanofiltration, *Desalination*, **219**(1), 171–178.

Nalwa H. (2000). Handbook of Nanostructured Materials and Nanotechnology, Vol. 5, Academic Press, San Diego, USA.

Namasivayam C. and Ranganathan K. (1995). Removal of Cd(II) from wastewater by adsorption on 'waste' Fe(III) Cr(III) hydroxide. *Water Research*, **29**(7), 1737–1744.

Ngomsik A., Bee A., Draye M., Cote G. and Cabuil V. (2005). Magnetic nano-and microparticles for metal removal and environmental applications: a review. *Comptes Rendus Chimie*, **8**(6), 963–970.

Nguyen C., Bang S., Cho J. and Kim K. (2009). Performance and mechanism of arsenic removal from water by a nanofiltration membrane. *Desalination*, **245**(1), 82–94.

Nightingale S., Worner H. and Dunne D. (1997). Microstructural development during the microwave sintering of yttria-zirconia ceramics. *Journal of the American Ceramic Society*, **80**(2), 394–400.

Nkegbe E. (2005). Assessment of heavy metal compositions in Glen Valley dry sludge, *Journal of Applied Science*, **5**(8), 1399–1401.

Nolan N., Synnott D., Seery M., Hinder S., Van Wassenhoven A. and Pillai, S. C. (2012). Effect of n-doping on the photocatalytic activity of sol-gel TiO_2. *Journal of Hazardous Materials*, **211**: 88–94.

Nordberg G., Goyer R. and Clarkson T. (1985). Impact of effects of acid precipitation on toxicity of metals. *Environmental Health Perspectives*, **63**: 169.

Nyquist R., Putzig C., Leugers M. and Kagel R. (1997). The Handbook of Infrared and Raman Spectra of Inorganic Compounds and Organic Salts: Infrared spectra of inorganic compounds (3800–4500 cm^{-1}), Vol. 4, Academic press.

Ohno T., Akiyoshi M., Umebayashi T., Asai K., Mitsui T. and Matsumura M. (2004). Preparation of S-doped TiO_2 photocatalysts and their photocatalytic activities under visible light. *Applied Catalysis A: General*, **265**(1), 115–121.

Okuyama K. and Wuled Lenggoro I. (2003). Preparation of nanoparticles via spray route. *Chemical Engineering Science*, **58**(3), 537–547.

Oladoja N., Ololade I., Olaseni S., Olatujoye V., Jegede O. and Agunloye, A. (2011). Synthesis of nano calcium oxide from a gastropod shell and the performance evaluation for Cr(VI) removal from aaqua system. *Industrial & Engineering Chemistry Research*, **51**(2), 639–648.

Ostroski I., Barros M., Silva E., Dantas J., Arroyo P. and Lima O. (2009). A comparative study for the ion exchange of Fe(III) and Zn(II) on zeolite NaY. *Journal of Hazardous Materials*, **161**(2), 1404–1412.

Pacheco S., Tapia J., Medina M. and Rodriguez R. (2006). Cadmium ions adsorption in simulated wastewater using structured alumina-silica nanoparticles. *Journal of Non-Crystalline Solids*, **352**(52), 5475–5481.

Pagenkopf G. (1983). Gill surface interaction model for trace-metal toxicity to fishes: role of complexation, pH, and water hardness. *Environmental Science & Technology*, **17**(6), 342–347.

Palchik O., Zhu J. and Gedanken A. (2000). Microwave assisted preparation of binary oxide nanoparticles. *Journal of Materials Chemistry*, **10**(5), 1251–1254.

Panatarani C., Lenggoro I. and Okuyama K. (2003). Synthesis of single crystalline ZnO nanoparticles by salt-assisted spray pyrolysis. *Journal of Nanoparticle Research*, **5**(1–2), 47–53.

Pang Y. L. and Abdullah A. (2013). Fe^{3+} doped TiO_2 nanotubes for combined adsorption-sonocatalytic degradation of real textile wastewater. *Applied Catalysis B: Environmental*, **129**, 473–481.

Pankove J. (1971). Optical Processes in Semiconductors. Dover Publications.

Panneerselvam M., Subanna G. and Rao K. (2001). Translucent yttrium aluminum garnet: microwave-assisted route to synthesis and processing. *Journal of Materials Research*, **16**(10), 2773–2776.

Panyala N., Peña-Méndez E. M. and Havel J. (2008). Silver or silver nanoparticles: a hazardous threat to the environment and human health. *Journal of Applied Biomedicine*, **6**(3), 117–29.

Park J., Min D. and Song H. (2002). Structural investigation of CaO-Al_2O_3 and CaO-Al_2O_3-CaF_2 slags via fourier transform infrared spectra. *The Iron and Steel Institute of Japan international*, **42**(1), 38–43.

Parker S. (2002). McGraw-Hill Dictionary of Scientific and Technical Terms. The McGraw-Hill Companies, New York.

Parks G. (1965). The isoelectric points of solid oxides, solid hydroxides, and aqueous hydroxo complex systems. *Chemical Reviews*, **65**(2), 177–198.

Peiró, A., Domingo C., Peral J., Domenech X., Vigil E., Hernández-Fenollosa M., Mollar M., Mari B. and Ayllon J. (2005). Nanostructured zinc oxide films grown from microwave activated aqueous solutions. *Thin Solid Films*, **483**(1), 79–83.

Penel G., Leroy G., Rey C., Sombret B., Huvenne J. and Bres E. (1997). Infrared and raman microspectrometry study of fluor-fluor-hydroxy and hydroxy-apatite powders. *Journal of Materials Science: Materials in Medicine*, **8**(5), 271–276.

Peng X., Li Y., Luan Z., Di Z., Wang H., Tian B. and Jia Z. (2003). Adsorption of 1, 2-dichlorobenzene from water to carbon nanotubes, *Chemical physics letters*, **376**(1), 154–158.

Peng Z. and Peng X. (2001). Mechanisms of the shape evolution of CdSe nanocrystals. *Journal of the American Chemical Society*, **123**(7), 1389–1395.

Peng Z. and Peng X. (2002). Nearly monodisperse and shape-controlled CdSe nanocrystals via alternative routes: nucleation and growth. *Journal of the American Chemical Society*, **124**(13), 3343–3353.

Peracchia M., Desmaële D., Couvreur P. and d'Angelo J. (1997). Synthesis of a novel poly (MePEG cyanoacrylate-co-alkyl cyanoacrylate) amphiphilic copolymer for nanoparticle technology. *Macromolecules*, **30**(4), 846–851.

Pérez-Maqueda L., Wang L. and Matijevic E. (1998). Nanosize indium hydroxide by peptization of colloidal precipitates, *Langmuir*, **14**(16), 4397–4401.

Pérez R., Cabrera G., Gómez J., Abalos A. and Cantero D. (2010). Combined strategy for the precipitation of heavy metals and biodegradation of petroleum in industrial wastewaters. *Journal of Hazardous Materials*, **182**(1), 896–902.

Pérez-Tijerina E., Pinilla M., Mejía-Rosales S., Ortiz-Méndez, U., Torres A. and José-Yacamán M. (2008). Highly size-controlled synthesis of Au/Pd nanoparticles by inert-gas condensation. *Faraday Discussions*, **138**: 353–362.

Peters R. and Bennett G. (1989). The simultaneous removal of oil and heavy metals from industrial wastewaters using hydroxide or sulfide precipitation coupled with air flotation. *Hazardous Waste and Hazardous Materials*, **6**(4), 327–345.

Peters R. and Ku Y. (1987). Removal of sulfides from waters and wastewaters by activated carbon. *Reactive Polymers, Ion Exchangers, Sorbents*, **5**(1), 93–104.

Pham H., McDowell T. and Wilkins E. (1995). Photocatalytically-mediated disinfection of water using TiO_2 as a catalyst and spore-forming *Bacillus pumilus* as a model. *Journal of Environmental Science & Health Part A*, **30**(3), 627–636.

Pierotti R. and Rouquerol J. (1985). Reporting physisorption data for gas/solid systems with special reference to the determination of surface area and porosity. *Pure and Applied Chemistry*, **57**(4), 603–619.

Pillay K., Cukrowska E. and Coville N. (2009). Multi-walled carbon nanotubes as adsorbents for the removal of parts per billion levels of hexavalent chromium from aqueous solution. *Journal of Hazardous Materials*, **166**(2–3), 1067–1075.

Plattes M., Bertrand A., Schmitt B., Sinner J., Verstraeten F. and Welfring J. (2007). Removal of tungsten oxyanions from industrial wastewater by precipitation, coagulation and flocculation processes. *Journal of Hazardous Materials*, **148**(3), 613–615.

Polat H. and Erdogan D. (2007). Heavy metal removal from waste waters by ion flotation. *Journal of Hazardous Materials*, **148**(1), 267–273.

Pozan G. and Kambur A. (2013). Removal of 4-chlorophenol from wastewater: preparation, characterization and photocatalytic activity of alkaline earth oxide doped TiO_2. *Applied Catalysis B: Environmental*, **129**: 409–415.

Qiu X., Li G., Sun X., Li L. and Fu X. (2008). Doping effects of Co^{2+} ions on ZnO nanorods and their photocatalytic properties. *Nanotechnology*, **19**(21), 215703.

Radoičić, M., Šaponjić, Z., Janković, I., Ćirić-Marjanović, G., Ahrenkiel S. and Comor M. (2013). Improvements to the photocatalytic efficiency of polyaniline modified TiO_2 nanoparticles. *Applied Catalysis B: Environmental*, **136**: 133–139.

Rai A., Prabhune A. and Perry C. (2010). Antibiotic mediated synthesis of gold nanoparticles with potent antimicrobial activity and their application in antimicrobial coatings. *Journal of Materials Chemistry*, **20**(32), 6789–6798.

Rai M., Yadav A. and Gade A. (2009). Silver nanoparticles as a new generation of antimicrobials. *Biotechnology Advances*, **27**(1), 76–83.

Rana S. (2006). Environmental Pollution: Health and Toxicology, Alpha Science Int'l Ltd.

Rao K., Vaidhyanathan B., Ganguli M. and Ramakrishnan P. (1999). Synthesis of inorganic solids using microwaves. *Chemistry of Materials*, **11**(4), 882–895.

Reddy K., Feris K., Bell J., Wingett D., Hanley C. and Punnoose A. (2007). Selective toxicity of zinc oxide nanoparticles to prokaryotic and eukaryotic systems. *Applied Physics Letters*, **90**(21), 213902.

Reddy S., Reddy K., Lakshminaraya S., Priya D., Rao Y. and Reddy A. (2008). Extractive spectrophotometric determination of trace amounts of cadmium (II) in medicinal leaves and environmental samples using benzildithiosemicarbazone (BDTSC). *Journal of Hazardous Materials* **152**(3), 903–909.

Rees G., Evans-Gowing R., Hammond S. and Robinson B. (1999). Formation and morphology of calcium sulfate nanoparticles and nanowires in water-in-oil microemulsions. *Langmuir*, **15**(6), 1993–2002.

Ren G., Hu D., Cheng E., Vargas-Reus M., Reip P. and Allaker R. (2009). Characterisation of copper oxide nanoparticles for antimicrobial applications. *International Journal of Antimicrobial Agents*, **33**(6), 587–590.

Reyes I., Villarroel M., Diez M. and Navia R. (2009). Using lignimerin (a recovered organic material from Kraft cellulose mill wastewater) as sorbent for Cu and Zn retention from aqueous solutions, *Bioresource Technology*, **100**(20), 4676–4682.

Richards R., Li W., Decker S., Davidson C., Koper O., Zaikovski V., Volodin A., Rieker T. and Klabunde K. (2000). Consolidation of metal oxide nanocrystals. reactive pellets with controllable pore structure that represent a new family of porous, inorganic materials. *Journal of the American Chemical Society*, **122**(20), 4921–4925.

Roy A. and Bhattacharya J. (2010). Synthesis of $Ca(OH)_2$ nanoparticles by wet chemical method. *Micro & Nano Letters*, **5**(2), 131–134.

Roy A. and Bhattacharya J. (2011). Microwave-assisted synthesis and characterization of CaO nanoparticles. *International Journal of Nanoscience*, **10**(03), 413–418.

Roy A. and Bhattacharya J. (2012a). Microwave-assisted synthesis and characterization of CaS nanoparticles, *International Journal of Nanoscience*, **11**(5), DOI: 10.1142/S0219581X12500275

Roy A. and Bhattacharya J. (2012b). Removal of Cu(II) Zn(II) and Pb(II) from water using microwave-assisted synthesized maghemite nanotubes. *Chemical Engineering Journal*, **211–212**, 493500.

Ruparelia J., Chatterjee A., Duttagupta S. and Mukherji S. (2008). Strain specificity in antimicrobial activity of silver and copper nanoparticles. *Acta Biomaterialia*, **4**(3), 707–716.

Ryskin Y. and Farmer V. (1974). The infrared spectra of minerals. *Mineralogical Society Monograph*, **4**, 137–182.

Sadiq I. M., Chowdhury B., Chandrasekaran N. and Mukherjee A. (2009). Antimicrobial sensitivity of *Escherichia coli* to alumina nanoparticles. *Nanomedicine: Nanotechnology, Biology and Medicine*, **5**(3), 282–286.

Sadler P., Higham D. and Nickolson J. (1985). Environmental Inorganic Chemistry. VCH Publishers Inc., Florida, chapter The environmental chemistry of metals with examples from studies of the speciation of cadmium., pp. 239–271.

Salam M., Makki M. and Abdelaal M. (2011). Preparation and characterization of multi-walled carbon nanotubes/chitosan nanocomposite and its application for the removal of heavy metals from aqueous solution. *Journal of Alloys and Compounds* **509**(5), 2582–2587.

Salipira K., Mamba B., Krause R., Malefetse T. and Durbach S. (2007). Carbon nanotubes and cyclodextrin polymers for removing organic pollutants from water. *Environmental Chemistry Letters*, **5**(1), 13–17.

Salvadori B. and Dei L. (2001). Synthesis of $Ca(OH)_2$ nanoparticles from diols. *Langmuir* **17**(8), 2371–2374.

Samper E., Rodríguez M., De la Rubia M. and Prats D. (2009). Removal of metal ions at low concentration by micellar-enhanced ultrafiltration (MEUF) using sodium dodecyl sulfate (SDS) and linear alkylbenzene sulfonate (LAS). *Separation and Purification Technology* **65**(3), 337–342.

Sanfeliu C., Sebastià, J., Cristòfol R. and Rodríguez-Farré, E. (2003). Neurotoxicity of organomercurial compounds. *Neurotoxicity Research*, **5**(4), 283–305.

Santos A., Alonso E., Callejon M. and Jimenez J. (2002). Heavy metal content and speciation in groundwater of the Guadiamar river basin. *Chemosphere*, **48**(3), 279–285.

Santos I., Silva-Filho E., Schaefer C., Albuquerque-Filho M. and Campos L. (2005). Heavy metal contamination in coastal sediments and soils near the Brazilian Antarctic Station, King George Island. *Marine Pollution Bulletin*, **50**(2), 185–194.

Saraf L., Patil S., Ogale S., Sainkar S. and Kshirsager S. (1998). Synthesis of nanophase tio_2 by ion beam sputtering and cold condensation technique. *International Journal of Modern Physics B*, **12**(25), 2635–2647.

Sarkar M., Chaudhuri G., Chattopadhyay A. and Biswas N. (2003). Effect of sodium arsenite on spermatogenesis, plasma gonadotrophins and testosterone in rats. *Asian Journal of Andrology*, **5**(1), 27–32.

Sauchelli V. (1969). Trace Elements in Agriculture, Van Nostrand Reinhold Co., New York, Cincinnati, Toronto, London, Melbourne.

Schierz A. and Zänker H. (2009). Aqueous suspensions of carbon nanotubes: surface oxidation, colloidal stability and uranium sorption. *Environmental Pollution*, **157**(4), 1088–1094.

Schnoor J. (1996). Environmental Modeling: Fate and Transport of Pollutants in Water, Air, and Soil. John Wiley and Sons.

Sengupta A. (2001). Environmental Separation of Heavy Metals: Engineering Processes, CRC Press.

Senthilnathan J. and Philip L. (2010). Photocatalytic degradation of lindane under UV and visible light using N-doped TiO_2. *Chemical Engineering Journal*, **161**(1), 83–92.

Shahalam A., Al-Harthy A. and Al-Zawhry A. (2002). Feed water pretreatment in RO systems: unit processes in the Middle East. *Desalination*, **150**(3), 235–245.

Shahverdi A., Fakhimi A., Shahverdi H. and Minaian S. (2007). Synthesis and effect of silver nanoparticles on the antibacterial activity of different antibiotics against *Staphylococcus aureus* and *Escherichia coli*. *Nanomedicine: Nanotechnology, Biology and Medicine*, **3**(2), 168–171.

Shanker G., Syversen T. and Aschner M. (2003). Astrocyte-mediated methylmercury neurotoxicity. *Biological Trace Element Research*, **95**(1), 1–10.

Shen X., Liu H., Pan L., Chen K., Hong J. and Xu Z. (2004). An efficient template pathway to synthesis of ordered metal oxide nanotube arrays using metal acetylacetonates as single-source molecular precursors. *Chemistry Letters*, **33**(9), 1128–1129.

Shen Y., Tang J., Nie Z., Wang Y., Ren Y. and Zuo L. (2009). Preparation and application of magnetic Fe_3O_4 nanoparticles for wastewater purification. *Separation and Purification Technology* **68**(3), 312–319.

Sheng G., Wang S., Hu J., Lu Y., Li J., Dong Y. and Wang X. (2009). Adsorption of Pb (II) on diatomite as affected via aqueous solution chemistry and temperature. *Colloids and Surfaces A: Physicochemical and Engineering Aspects*, **339**(1), 159–166.

Singh S., Barick K. and Bahadur D. (2011a). Novel and efficient three dimensional mesoporous ZnO nanoassemblies for envirnomental remediation. *International Journal of Nanoscience* **10**(04n05), 1001–1005.

Singh S., Barick K. and Bahadur D. (2011b). Surface engineered magnetic nanoparticles for removal of toxic metal ions and bacterial pathogens. *Journal of Hazardous Materials*, **192**(3), 1539–1547.

Singh S., Barick K. and Bahadur D. (2013). Shape-controlled hierarchical ZnO architectures: photocatalytic and antibacterial activities. *CrystEngComm*, **15**(23), 4631–4639.

Skrabalak S. and Suslick K. (2005). Porous mos_2 synthesized by ultrasonic spray pyrolysis. *Journal of the American Chemical Society*, **127**(28), 9990–9991.

Smara A., Delimi R., Chainet E. and Sandeaux J. (2007). Removal of heavy metals from diluted mixtures by a hybrid ion-exchange/electrodialysis process. *Separation and Purification Technology*, **57**(1), 103–110.

Smeitink J., Sengers R., Trijbels F. and van den Heuvel L. (2000). Nuclear genes and oxidative phosphorylation disorders: a review. *European Journal of Pediatrics*, **159**(3), S227–S231.

Smit J. and Wijn H. (1959). Ferrites. Philips Technical Library, Eindhoven.

Society R. and of Engineering R. A. (2004). Nanoscience and nanotechnologies: opportunities and uncertainties, p. 113.

Sondi I. and Salopek-Sondi B. (2004). Silver nanoparticles as antimicrobial agent: a case study on *E. coli* as a model for gram-negative bacteria. *Journal of Colloid and Iinterface Science*, **275**(1), 177–182.

Sooklal K., Cullum B., Angel S. and Murphy C. (1996). Photophysical properties of ZnS nanoclusters with spatially localized Mn^{2+}. *The Journal of Physical Chemistry*, **100**(11), 4551–4555.

Srivastava N. and Majumder C. (2008). Novel biofiltration methods for the treatment of heavy metals from industrial wastewater. *Journal of Hazardous Materials*, **151**(1), 1–8.

Stephenson R. and Blackburn J. (1998). The Industrial Wastewater Systems: Handbook, CRC PressI Llc.

Stoimenov P., Klinger R., Marchin G. and Klabunde K. (2002). Metal oxide nanoparticles as bactericidal agents. *Langmuir*, **18**(17), 6679–6686.

Sugimoto T. and Matijević E. (1980). Formation of uniform spherical magnetite particles by crystallization from ferrous hydroxide gels. *Journal of Colloid and Interface Science* **74**(1), 227–243.

Sui Y., Skomski R., Sorge K. and Sellmyer D. (2004). Magnetic nanotubes produced by hydrogen reduction. *Journal of Applied Physics*, **95**: 7151–7153.

Sun Z., Yuan H., Liu Z., Han B. and Zhang X. (2005). A highly efficient chemical sensor material for H_2S: α-Fe_2O_3 nanotubes fabricated using carbon nanotube templates, *Advanced Materials*, **17**(24), 2993–2997.

Sutton W. (1989). Microwave processing of ceramic materials. *American Ceramic Society Bulletin*, **68**(2), 376–386.

Sze S. (2008). Semiconductor Devices: Physics and Technology, John Wiley & Sons.

Taffarel S. and Rubio J. (2009). On the removal of Mn^{2+} ions by adsorption onto natural and activated Chilean zeolites. *Minerals Engineering*, **22**(4), 336–343.

Talapin D., Shevchenko E., Murray C., Titov A. and Král P. (2007). Dipole-dipole interactions in nanoparticle superlattices. *Nano Letters*, **7**(5), 1213–1219.

Tanaka S. (1988). Color electroluminescence in alkaline-earth sulfide thin-films, *Journal of Luminescence*, **40**, 20–23.

Taniguchi N. (1974). On the basic concept of nanotechnology, *Proc. Intl. Conf. Prod. Eng. Tokyo, Part II, Japan Society of Precision Engineering*, pp. 18–23.

Tassel F., Rubio J., Misra M. and Jena B. (1997). Removal of mercury from gold cyanide solution by dissolved air flotation. *Minerals Engineering*, **10**(8), 803–811.

Tavlarides L., Bae J. and Lee C. (1987). Solvent extraction, membranes, and ion exchange in hydrometallurgical dilute metals separation. *Separation Science and Technology*, **22**(2–3), 581–617.

Tekaia-Elhsissen K., Delahaye-Vidal A., Nowogrocki G. and Figlarz M. (1989). Chimie de l'état solide: réaction de l'hydroxyde de nickel avec l'éthylèneglycol. *Comptes rendus de l'Académie des sciences. Série 2, Mécanique, Physique, Chimie, Sciences de l'univers, Sciences de la Terre*, **309**(4), 349–352.

Ter Maat H., Hogendoorn J. and Versteeg G. (2005). The removal of hydrogen sulfide from gas streams using an aqueous metal sulfate absorbent: Part I. The absorption of hydrogen sulfide in metal sulfate solutions. *Separation and Purification Technology*, **43**(3), 183–197.

Tessele F., Misra M. and Rubio J. (1998). Removal of Hg, As and Se ions from gold cyanide leach solutions by dissolved air flotation. *Minerals Engineering*, **11**(6), 535–543.

Thill A., Zeyons O., Spalla O., Chauvat F., Rose J., Auffan M. and Flank, A. (2006). Cytotoxicity of CeO_2 nanoparticles for *Escherichia coli* physico-chemical insight of the cytotoxicity mechanism. *Environmental Science & Technology*, **40**(19), 6151–6156.

Torimoto T., Okazaki K., Kiyama T., Hirahara K., Tanaka N. and Kuwabata, S. (2006). Sputter deposition onto ionic liquids: simple and clean synthesis of highly dispersed ultrafine metal nanoparticles. *Applied Physics Letters* **89**(24), 243117.

Toze S. (2006). Reuse of effluent water-benefits and risks. *Agricultural Water Management*, **80**(1), 147–159.

Ui J. (1992). Industrial Pollution in Japan, United Nations University Press Tokyo.

Ullah R. and Dutta J. (2008). Photocatalytic degradation of organic dyes with manganese-doped ZnO nanoparticles. *Journal of Hazardous Materials*, **156**(1), 194–200.

USEPA (1999). Development Document for Effluent Limitations Guidelines and Standards for the Metal Finishing Point Source Category, *Technical report*, US EPA, Washington, DC.

USEPA (2011). The Drinking Water Standards and Health Advisories. Technical Report EPA 820-R-11-002, United States Environmental Protection Agency.

Veeken A., Akoto L., Hulshoff Pol L. and Weijma J. (2003). Control of the sulfide (S^{2-}) concentration for optimal zinc removal by sulfide precipitation in a continuously stirred tank reactor. *Water Research*, **37**(15), 3709–3717.

Vieira R. and Beppu M. (2006). Dynamic and static adsorption and desorption of Hg (II) ions on chitosan membranes and spheres. *Water Research*, **40**(8), 1726–1734.

Vogel R., Hoyer P. and Weller H. (1994). Quantum-sized PbS, CdS, Ag_2S, Sb_2S_3, and Bi_2S_3 particles as sensitizers for various nanoporous wide-bandgap semiconductors. *The Journal of Physical Chemistry*, **98**(12), 3183–3188.

Volesky B. (1990). Removal and Recovery of Heavy Metals by Biosorption. Vol. 7. CRC Press, Boca Raton, FL.

Vuković, G., Marinković, A., Čolić, M., Ristić, M., Aleksić, R., Perić-Grujić, A. and Uskoković, P. (2010). Removal of cadmium from aqueous solutions by oxidized and ethylenediamine-functionalized multi-walled carbon nanotubes. *Chemical Engineering Journal*, **157**(1), 238–248.

Vuković, G., Marinković, A., Škapin S., Ristić, M., Aleksić, R., Perić-Grujić, A. and Uskoković, P. (2011). Removal of lead from water by amino modified multi-walled carbon nanotubes. *Chemical Engineering Journal*, **173**(3), 855–865.

Wagner G., Koper O., Lucas E., Decker S. and Klabunde K. (2000). Reactions of VX, GD, and HD with nanosize CaO: autocatalytic dehydrohalogenation of HD. *The Journal of Physical Chemistry B*, **104**(21), 5118–5123.

Walton F., Harmon A., Paul D., Drobná, Z., Patel Y. and Styblo M. (2004). Inhibition of insulin-dependent glucose uptake by trivalent arsenicals: possible mechanism of arsenic-induced diabetes. *Toxicology and Applied Pharmacology*, **198**(3), 424–433.

Wan J. (1993). Microwaves and chemistry: the catalysis of an exciting marriage. *Research on Chemical Intermediates*, **19**(2), 147–158.

Wan N., Endud C. and Mayanar R. (2002). Removal of copper (II) ions from aqueous solution onto chitosan and cross-linked chitosan beads. *Reactive and Functional Polymers*, **50**(2), 181–190.

Wang H., Zhou A., Peng F., Yu H. and Yang J. (2007). Mechanism study on adsorption of acidified multiwalled carbon nanotubes to Pb (II). *Journal of Colloid and Iinterface Science*, **316**(2), 277–283.

Wang H., Zhu J., Zhu J., Liao X., Xu S., Ding T. and Chen H. (2002). Preparation of nanocrystalline ceria particles by sonochemical and microwave assisted heating methods. *Physical Chemistry Chemical Physics* **4**(15), 3794–3799.

Wang J. X., Jiang D. Q., Gu Z. Y. and Yan X. P. (2006a). Multiwalled carbon nanotubes coated fibers for solid-phase microextraction of polybrominated diphenyl ethers in water and milk samples before gas chromatography with electron-capture detection. *Journal of chromatography A*, **1137**(1), 8–14.

Wang L., Li J., Jiang Q. and Zhao L. (2012). Water-soluble Fe_3O_4 nanoparticles with high solubility for removal of heavy-metal ions from waste water. *Dalton Transactions*, **41**(15), 4544–4551.

Wang S., Gu F., Lü, M., Song C., Liu S., Xu D. and Yuan D. (2003). Preparation and characterization of sol-gel derived $ZnTiO_3$ nanocrystals. *Materials Research Bulletin*, **38**(8), 1283–1288.

Wang T., Wang Y., Li F., Xu C. and Zhou D. (2006b). Morphology and magnetic behaviour of an Fe_3O_4 nanotube array, *Journal of Physics: Condensed Matter*, **18**: 10545.

Wang Y., Feng C., Zhang M., Yang J. and Zhang Z. (2010). Enhanced visible light photocatalytic activity of N-doped TiO_2 in relation to single-electron-trapped oxygen vacancy and doped-nitrogen, *Applied Catalysis B: Environmental*, **100**(1), 84–90.

Waters A. (1990). Dissolved air flotation used as primary separation for heavy metal removal. *Filtration & Separation*, **27**(2), 70–73.

Webb J. (2004). Use of the ecosystem approach to population health: the case of mercury contamination in aquatic environments and riparian populations, Andean Amazon, Napo River Valley, Ecuador.. *Canadian journal of public health. Revue canadienne de sante publique*, **96**(1), 44–46.

Wei C., Lin W., Zainal Z., Williams N., Zhu K., Kruzic A., Smith R. L. and Rajeshwar K. (1994). Bactericidal activity of TiO_2 photocatalyst in aqueous media: toward a solar-assisted water disinfection system. *Environmental Science & Technology*, **28**(5), 934–938.

Wei X., Viadero Jr R. and Buzby K. (2005). Recovery of iron and aluminum from acid mine drainage by selective precipitation. *Environmental Engineering Science*, **22**(6), 745–755.

Weiss B. (1995). Long ago and far away: a retrospective on the implications of Minamata. *Neurotoxicology*, **17**(1), 257–263.

Welz B. and Sperling M. (2008). Atomic Absorption Spectrometry, John Wiley & Sons.

WHO (1993). Guidelines for Drinking Water Quality (2nd ed.), Technical report, World Health Organization, Geneva.

Wiesbrock F., Hoogenboom R., Abeln C. and Schubert U. (2004). Single-mode microwave ovens as new reaction devices: accelerating the living polymerization of 2-Ethyl-2-Oxazoline. *Macromolecular Rapid Communications*, **25**(22), 1895–1899.

Wigginton N., Haus K. and Hochella Jr M. (2007). Aquatic environmental nanoparticles. *Journal of Environmental Monitoring*, **9**(12), 1306–1316.

Wilhelmy D. and Matijević, E. (1984). Preparation and properties of monodispersed spherical-colloidal particles of zinc sulphide. *Journal of the Chemical Society, Faraday Transactions 1: Physical Chemistry in Condensed Phases*, **80**(3), 563–570.

Wong S., Li X., Zhang G., Qi S. and Min Y. (2002). Heavy metals in agricultural soils of the Pearl River Delta, South China. *Environmental Pollution*, **119**(1), 33–44.

Wu B. and Kumar A. (2007). Extreme ultraviolet lithography: A Review. *Journal of Vacuum Science & Technology B: Microelectronics and Nanometer Structures*, **25**(6), 1743–1761.

Wu L. and Ritchie S. (2008). Enhanced dechlorination of trichloroethylene by membrane-supported Pd-coated iron nanoparticles. *Environmental Progress*, **27**(2), 218–224.

Wu L., Shamsuzzoha M. and Ritchie S. (2005a). Preparation of cellulose acetate supported zero-valent iron nanoparticles for the dechlorination of trichloroethylene in water. *Journal of Nanoparticle Research*, **7**(4–5), 469–476.

Wu P., Wang W., Huang Y., Sheu H., Lo Y., Tsai T., Shieh D. and Yeh C. (2007). Porous iron oxide based nanorods developed as delivery nanocapsules. *Chemistry-A European Journal*, **13**(14), 3878–3885.

Wu R., Qu J. and Chen Y. (2005b). Magnetic powder MnO-Fe_2O_3 composite-a novel material for the removal of azo-dye from water. *Water Research*, **39**(4), 630–638.

Wu W., Xiao X., Zhang S., Peng T., Zhou J., Ren F. and Jiang C. (2010). Synthesis and magnetic properties of maghemite (γ-Fe_2O_3) short-nanotubes. *Nanoscale Research Letters* **5**(9), 1474–1479.

Wu Z. and Gao J. (2012). Synthesis of γ-Fe_2O_3 nanoparticles by homogeneous co-precipitation method. *Micro & Nano Letters,*, **7**(6), 533–535.

Wu Z., Dong F., Zhao W., Wang H., Liu Y. and Guan B. (2009). The fabrication and characterization of novel carbon doped TiO_2 nanotubes, nanowires and nanorods with high visible light photocatalytic activity. *Nanotechnology*, **20**(23), 235701.

Xiaolin S., Guangyan H., Guilan Z., Guoqing T. and Wenju C. (2001). The luminescence properties of CaS: Eu^{2+} excited by picosecond laser pulse. *Journal of luminescence*, **92**(4), 307–310.

Xie M., Jing L., Zhou J., Lin J. and Fu H. (2010). Synthesis of nanocrystalline anatase TiO_2 by one-pot two-phase separated hydrolysis-solvothermal processes and its high activity for photocatalytic degradation of rhodamine B. *Journal of Hazardous Materials*, **176**(1), 139–145.

Xiong Y., Xie Y., Li Z., Zhang R., Yang J. and Wu C. (2003). Complexing-reagent assisted synthesis of α-Fe and γ-Fe_2O_3 nanowires under mild conditions. *New Journal of Chemistry*, **27**(3), 588–590.

Xu L., Hu Y., Pelligra C., Chen C., Jin L., Huang H., Sithambaram S., Aindow M., Joesten R. and Suib S. (2009). ZnO with different morphologies synthesized by solvothermal methods for enhanced photocatalytic activity. *Chemistry of Materials*, **21**(13), 2875–2885.

Yamanaka M., Hara K. and Kudo J. (2005). Bactericidal actions of a silver ion solution on *Escherichia coli*, studied by energy-filtering transmission electron microscopy and proteomic analysis. *Applied and Environmental Microbiology*, **71**(11), 7589–7593.

Yamato N. (1988). Concentrations and chemical species of arsenic in human urine and hair. *Bulletin of Environmental Contamination and Toxicology*, **40**(5), 633–640.

Yang J., Lin C., Wang Z. and Lin J. (2006a). $In(OH)_3$ and In_2O_3 nanorod bundles and spheres: microemulsion-mediated hydrothermal synthesis and luminescence properties. *Inorganic Chemistry*, **45**(22), 8973–8979.

Yang K. and Xing B. (2007). Desorption of polycyclic aromatic hydrocarbons from carbon nanomaterials in water. *Environmental Pollution*, **145**(2), 529–537.

Yang K., Wang X. L., Zhu L. Z. and Xing B. (2006b). Competitive sorption of polycyclic aromatic hydrocarbons on carbon nanotubes. *Environmental science & technology*, **40**: 5804–5810.

Yang K., Zhu L. and Xing B. (2006c). Adsorption of polycyclic aromatic hydrocarbons by carbon nanomaterials. *Environmental science & technology*, **40**(6), 1855–1861.

Yang L., Liu P., Li X. and Li S. (2012). The photo-catalytic activities of neodymium and fluorine doped TiO_2 nanoparticles. *Ceramics International*, **38**(6), 4791–4796.

Yang P., Lü, M., Xü, D., Yuan D., Chang J., Zhou G. and Pan M. (2002). Strong green luminescence of Ni^{2+}-doped ZnS nanocrystals. *Applied Physics A: Materials Science & Processing*, **74**(2), 257–259.

Yang X., Cao C., Erickson L., Hohn K., Maghirang R. and Klabunde K. (2009). Photo-catalytic degradation of rhodamine B on C-, S-, N-, and Fe-doped TiO_2 under visible-light irradiation. *Applied Catalysis B: Environmental*, **91**(3), 657–662.

Yang Y. and Wang P. (2006). Preparation and characterizations of a new PS/TiO_2 hybrid membranes by sol-gel process. *Polymer*, **47**(8), 2683–2688.

Yantasee W., Warner C., Sangvanich T., Addleman R., Carter T., Wiacek R., Fryxell G., Timchalk C. and Warner M. (2007). Removal of heavy metals from aqueous systems with thiol functionalized superparamagnetic nanoparticles. *Environmental science & technology*, **41**(14), 5114–5119.

Yavuz C., Mayo J., William W., Prakash A., Falkner J., Yean S., Cong L., Shipley H., Kan A. and Tomson M. (2006a). Low-field magnetic separation of monodisperse Fe_3O_4 nanocrystals. *Science*, **314**(5801), 964–967.

Yavuz R., Orbak, İ. and Karatepe N. (2006b). Factors affecting the adsorption of chromium (VI) on activated carbon. *Journal of Environmental Science and Health Part A*, **41**(9), 1967–1980.

Yıldırım, Ö., Unalan H. and Durucan C. (2013). Highly efficient room temperature synthesis of silver-doped zinc oxide (ZnO: Ag) nanoparticles: structural, optical, and photocatalytic properties. *Journal of the American Ceramic Society*, **96**(3), 766–773.

Yin S., Luo Z., Xia J. and Li H. (2010). Microwave-assisted synthesis of Fe_3O_4 nanorods and nanowires in an ionic liquid. *Journal of Physics and Chemistry of Solids*, **71**(12), 1785–1788.

Yoshida T., Yamauchi H. and Fan Sun G. (2004). Chronic health effects in people exposed to arsenic via the drinking water: dose-response relationships in review. *Toxicology and Applied Pharmacology*, **198**(3), 243–252.

Yu I., Isobe T. and Senna M. (1996). Optical properties and characteristics of ZnS nano-particles with homogeneous Mn distribution. *Journal of Physics and Chemistry of Solids*, **57**(4), 373–379.

Yu J., Dai G., Xiang Q. and Jaroniec M. (2011). Fabrication and enhanced visible-light photocatalytic activity of carbon self-doped TiO_2 sheets with exposed {001} facets. *Journal of Materials Chemistry*, **21**(4), 1049–1057.

Yuan G. and Wu L. (2007). Allophane nanoclay for the removal of phosphorus in water and wastewater. *Science and Technology of Advanced Materials*, **8**(1), 60–62.

Yuan X., Meng Y., Zeng G., Fang Y. and Shi J. (2008). Evaluation of tea-derived biosurfactant on removing heavy metal ions from dilute wastewater by ion flotation. *Colloids and Surfaces A: Physicochemical and Engineering Aspects*, **317**(1), 256–261.

Yura K., Fredrikson K. and Matijević, E. (1990). Preparation and properties of uniform colloidal indium compounds of different morphologies. *Colloids and Surfaces*, **50**: 281–293.

Zamboulis D., Pataroudi S., Zouboulis A. and Matis K. (2004). The application of sorptive flotation for the removal of metal ions, *Desalination*, **162**: 159–168.

Zeng H., Li J., Liu J., Wang Z. and Sun S. (2002). Exchange-coupled nanocomposite magnets by nanoparticle self-assembly. *Nature*, **420**(6914), 395–398.

Zhai T., Xie S., Zhao Y., Sun X., Lu X., Yu M., Xu M., Xiao F. and Tong Y. (2012). Controllable synthesis of hierarchical ZnO nanodisks for highly photocatalytic activity. *CrystEngComm*, **14**(5), 1850–1855.

Zhang H., Chen X., Li Z., Kou J., Yu T. and Zou Z. (2007a). Preparation of sensitized ZnS and its photocatalytic activity under visible light irradiation. *Journal of Physics D: Applied Physics*, **40**(21), 6846.

Zhang H., Quan X., Chen S., Zhao H. and Zhao Y. (2006a). Fabrication of photocatalytic membrane and evaluation its efficiency in removal of organic pollutants from water. *Separation and Purification Technology*, **50**(2), 147–155.

Zhang H., Quan X., Chen S., Zhao H. and Zhao Y. (2006b). The removal of sodium dodecylbenzene sulfonate surfactant from water using silica/titania nanorods/nanotubes composite membrane with photocatalytic capability. *Applied Surface Science*, **252**(24), 8598–8604.

Zhang J., Wang Y., Ji H., Wei Y., Wu N., Zuo B. and Wang Q. (2005). Magnetic nanocomposite catalysts with high activity and selectivity for selective hydrogenation of *ortho*-chloronitrobenzene. *Journal of Catalysis*, **229**(1), 114–118.

Zhang L., Jiang Y., Ding Y., Povey M. and York D. (2007b). Investigation into the antibacterial behaviour of suspensions of ZnO nanoparticles (ZnO nanofluids). *Journal of Nanoparticle Research*, **9**(3), 479–489.

Zhang Q., Luo J., Vileno E. and Suib S. (1997). Synthesis of cryptomelane-type manganese oxides by microwave heating. *Chemistry of Materials*, **9**(10), 2090–2095.

Zhang S., Cheng F., Tao Z., Gao F. and Chen J. (2006c). Removal of nickel ions from wastewater by $Mg(OH)_2/MgO$ nanostructures embedded in Al_2O_3 membranes. *Journal of Alloys and Compounds*, **426**(1), 281–285.

Zhang Y., Li G., Jin Y., Zhang Y., Zhang J. and Zhang L. (2002). Hydrothermal synthesis and photoluminescence of tio_2 nanowires, *Chemical Physics Letters*, **365**(3), 300–304.

Zhao Y., Shen H., Pan S. and Hu M. (2010). Synthesis, characterization and properties of ethylenediamine-functionalized Fe_3O_4 magnetic polymers for removal of Cr(VI) in wastewater. *Journal of Hazardous Materials*, **182**(1), 295–302.

Zheng J., Jiang Z., Kuang Q., Xie Z., Huang R. and Zheng L. (2009). Shape-controlled fabrication of porous ZnO architectures and their photocatalytic properties. *Journal of Solid State Chemistry*, **182**(1), 115–121.

Zheng J., Liu Z., Liu X., Yan X., Li D. and Chu W. (2011). Facile hydrothermal synthesis and characteristics of B-doped TiO_2 hybrid hollow microspheres with higher photo-catalytic activity, *Journal of Alloys and Compounds*, **509**(9), 3771–3776.

Zhong L., Hu J., Liang H., Cao A., Song W. and Wan L. (2006). Self-assembled 3D flowerlike iron oxide nanostructures and their application in water treatment. *Advanced Materials*, **18**(18), 2426–2431.

Zhou Q., Ding Y. and Xiao J. (2006a). Sensitive determination of thiamethoxam, imidacloprid and acetamiprid in environmental water samples with solid-phase extraction packed with multiwalled carbon nanotubes prior to high-performance liquid chromatography. *Analytical and Bioanalytical Chemistry*, **385**(8), 1520–1525.

Zhou Q., Wang W. and Xiao J. (2006b). Preconcentration and determination of nicosulfuron, thifensulfuron-methyl and metsulfuron-methyl in water samples using carbon nanotubes packed cartridge in combination with high performance liquid chromatography. *Analytica Chimica Acta*, **559**(2), 200–206.

Zhou Q., Xiao J. and Wang W. (2006c). Using multi-walled carbon nanotubes as solid phase extraction adsorbents to determine dichlorodiphenyltrichloroethane and its metabolites at trace level in water samples by high performance liquid chromatography with UV detection. *Journal of chromatography A*, **1125**(2), 152–158.

Zhou Q., Xiao J. and Wang W. (2007). Trace analysis of triasulfuron and bensulfuron-methyl in water samples using a carbon nanotubes packed cartridge in combination with high-performance liquid chromatography. *Microchimica Acta*, **157**(1–2), 93–98.

Zhou Q., Xiao J., Wang W., Liu G., Shi Q. and Wang J. (2006d). Determination of atrazine and simazine in environmental water samples using multiwalled carbon nanotubes as the adsorbents for preconcentration prior to high performance liquid chromatography with diode array detector, *Talanta*, **68**(4), 1309–1315.

Zhu J., Aruna S., Koltypin Y. and Gedanken A. (2000). A novel method for the preparation of lead selenide: pulse sonoelectrochemical synthesis of lead selenide nanoparticles. *Chemistry of Materials*, **12**(1), 143–147.

Zhu J., Xu S., Wang H., Zhu J. and Chen H. (2003). Sonochemical synthesis of CdSe hollow spherical assemblies via an in-situ template route. *Advanced Materials*, **15**(2), 156–159.

Zhu S., Zhou H., Hibino M., Honma I. and Ichihara M. (2005). Synthesis of mno_2 nanoparticles confined in ordered mesoporous carbon using a sonochemical method. *Advanced Functional Materials*, **15**(3), 381–386.

Index